# The Definitive Guide to Google Vertex AI

Accelerate your machine learning journey with
Google Cloud Vertex AI and MLOps best practices

**Jasmeet Bhatia**

**Kartik Chaudhary**

BIRMINGHAM – MUMBAI

# The Definitive Guide to Google Vertex AI

**Group Product Manager**: Niranjan Naikwadi

**Publishing Product Manager**: Sanjana Gupta

**Book Project Manager**: Hemangi Lotlikar

**Senior Editor**: Gowri Rekha

**Technical Editor**: Rahul Limbachiya

**Copy Editor**: Safis Editing

**Project Coordinator**: Shambhavi Mishra

**Proofreader**: Safis Editing

**Indexer**: Pratik Shirodkar

**Production Designer**: Shankar Kalbhor

**DevRel Marketing Executive (DRME)**: Vinishka Kalra

First published: December 2023

Production reference: 1211223

Published by
Packt Publishing Ltd.
Grosvenor House
11 St Paul's Square
Birmingham
B3 1RB, UK

ISBN 978-1-80181-526-0

www.packtpub.com

*To my incredible wife, who gracefully navigates our cloud rivalry with love and patience. This book wouldn't be possible without your constant support and encouragement. Thank you for tolerating my late-night writing and even later-night snacking. Your patience is my favorite superpower.*

*To my parents, who still can't fully explain what I do for a living but are proud nonetheless – you're the original algorithms of my life. Thank you for programming me with constant love, support, and the occasional necessary reboot!*

*To my wonderful daughters, without whom I would never have really understood why so many authors joke about their kids delaying their books. Now I do. Thank you for bringing immense joy and well-needed system shutdowns to my life.*

*And to my colleagues, the wizards of Google Cloud, who speak fluent Python and dream in code – without you, this book would just be a collection of funny error messages.*

*This book is dedicated to all of you. May our models always converge, and may we all never run out of GPUs!*

*– Jasmeet Bhatia*

*To my mother, Smt. Sarita Devi, and my father, Mr. Inderpal Singh, for their sacrifices, constant love, and never-ending support. Thank you for teaching me to believe in myself, in God, and in my dreams.*

*To my little brother, Chakit Gill, for continuous encouragement, support, and love. Thanks for being my best friend; I am really proud of you.*

*To my friends and colleagues for their inspiration, motivation, and always being there for me.*

*And, most importantly, to all the readers – I hope this book helps you with your goals, because that's the real motivation behind writing this book and every single technical article that I share publicly on my blog.*

*– Kartik Chaudhary*

# Contributors

## About the authors

**Jasmeet Bhatia** is a machine learning solution architect with over 18 years of industry experience, with the last 10 years focused on global-scale data analytics and machine learning solutions. In his current role at Google, he works closely with key GCP enterprise customers to provide them guidance on how to best use Google's cutting-edge machine learning products. At Google, he has also worked as part of the Area 120 incubator on building innovative data products such as Demand Signals, and he has been involved in the launch of Google products such as Time Series Insights. Before Google, he worked in similar roles at Microsoft and Deloitte.

When not immersed in technology, he loves spending time with his wife and two daughters, reading books, watching movies, and exploring the scenic trails of southern California.

He holds a bachelor's degree in electronics engineering from Jamia Millia Islamia University in India and an MBA from the University of California Los Angeles (UCLA) Anderson School of Management.


**Kartik Chaudhary** is an AI enthusiast, educator, and ML professional with 6+ years of industry experience. He currently works as a senior AI engineer with Google to design and architect ML solutions for Google's strategic customers, leveraging core Google products, frameworks, and AI tools. He previously worked with UHG, as a data scientist, and helped in making the healthcare system work better for everyone. Kartik has filed nine patents at the intersection of AI and healthcare.

Kartik loves sharing knowledge and runs his own blog on AI, titled *Drops of AI*.

Away from work, he loves watching anime and movies and capturing the beauty of sunsets.

*I would like to thank my parents, my brother, and my friends for their constant love and support.*

# About the reviewers

**Surya Tripathi** is a seasoned data scientist with nearly nine years of expertise in data science, analysis, and data engineering. He holds a bachelor's degree in electronics and communications engineering and a master's in applied mathematics from Liverpool John Moores University. He is proficient in cloud platforms (GCP, Azure, AWS, and IBM) and has extensive GCP experience, delivering ML solutions in CPG, healthcare, banking, and supply chain. Involved in the full data science life cycle, he excels in requirement gathering, data analysis, model development, and MLOps. With experience in both consulting and product companies, he is currently affiliated with a top consulting firm, where his primary focus areas include generative AI and demand forecasting.

**Gopala Dhar** has worked with and implemented state-of-the-art technology in the field of AI to solve real-world business use cases at scale. He has four published patents to his name, ranging from the field of software design to hardware manufacturing, including embedded systems. His latest stint is at Google as an AI engineer. His areas of expertise include ML, ML system design, reinforcement learning, and, most recently, generative AI. He shares what he learns frequently through blog posts and open source contributions. He has won several awards from various academic as well as professional institutions, including the Indian Institute of Technology in Mumbai, the Indian Institute of Management in Bangalore, Texas Instruments, and Google.

**Lakshmanan Sethu Sankaranarayanan** is an award-winning AI/ML cloud industry leader in the fields of data, AI, and ML. He helps enterprise customers to migrate to Google Cloud, Azure, and AWS. He serves on the Technical Advisory Board for AI/ML solutions at Packt, and he is also the Technical Editor for Packt and O'Reilly. He has been honored with three LinkedIn TopVoice awards for his contributions to AI/ML and cloud computing. He earned four Microsoft Most Valuable Player awards for his outstanding contribution to the cloud community.

*I would like to acknowledge my wife, Dhivya, and my kids, Sanjana and Saisasthik, for being a constant source of support and encouragement throughout this book-reviewing journey.*

**Chetan Apsunde** is an experienced software engineer, specializing in conversational AI and machine learning with a robust nine-year IT background. He works with Google to build cloud solutions using CCAI and GenAI. He is passionate about creating intelligent, user-centric solutions at the intersection of technology and human interaction.

# Table of Contents

**Preface**                                                                    **xv**

**Part 1: The Importance of MLOps in a Real-World ML Deployment**

1

**Machine Learning Project Life Cycle and Challenges**                         **3**

| | | | |
|---|---|---|---|
| ML project life cycle | 3 | Infrastructure requirements | 12 |
| Common challenges in developing real-world ML solutions | 9 | **Limitations of ML** | **12** |
| Data collection and security | 10 | Data-related concerns | 13 |
| Non-representative training data | 10 | Deterministic nature of problems | 13 |
| Poor quality of data | 11 | Lack of interpretability and reproducibility | 14 |
| Underfitting the training dataset | 11 | Concerns related to cost and customizations | 14 |
| Overfitting the training dataset | 11 | Ethical concerns and bias | 15 |
| | | **Summary** | **15** |

2

**What Is MLOps, and Why Is It So Important for Every ML Team?**              **17**

| | | | |
|---|---|---|---|
| Why is MLOps important? | 18 | MLOps maturity level 2 – automated model deployments | 22 |
| Implementing different MLOps maturity levels | 19 | How can Vertex AI help with implementing MLOps? | 24 |
| MLOps maturity level 0 | 19 | **Summary** | **34** |
| MLOps maturity level 1 – automating basic ML steps | 20 | | |

# Part 2: Machine Learning Tools for Custom Models on Google Cloud

## 3

## It's All About Data – Options to Store and Transform  ML Datasets 37

| | | | |
|---|---|---|---|
| Moving data to Google Cloud | 38 | Transforming data | 43 |
| Google Cloud Storage Transfer tools | 38 | Ad hoc transformations within Jupyter | |
| BigQuery Data Transfer Service | 40 | Notebook | 43 |
| Storage Transfer Service | 40 | Cloud Data Fusion | 46 |
| Transfer Appliance | 40 | Dataflow pipelines for scalable data | |
| | | transformations | 47 |
| Where to store data | 41 | Summary | 47 |
| GCS – object storage | 41 | | |
| BQ – data warehouse | 42 | | |

## 4

## Vertex AI Workbench – a One-Stop Tool for AI/ML Development Needs 49

| | | | |
|---|---|---|---|
| What is Jupyter Notebook? | 50 | Custom containers for Vertex AI | |
| Getting started with Jupyter Notebook | 50 | Workbench | 58 |
| Vertex AI Workbench | 52 | Scheduling notebooks in Vertex AI | 59 |
| Getting started with Vertex AI Workbench | 53 | Configuring notebook executions | 59 |
| | | Summary | 63 |

## 5

## No-Code Options for Building ML Models 65

| | | | |
|---|---|---|---|
| ML modeling options in Google Cloud | 66 | Importing data to use with Vertex AI AutoML | 70 |
| What is AutoML? | 68 | Training the AutoML model for | |
| Vertex AI AutoML | 69 | tabular/structured data | 76 |
| How to create a Vertex AI AutoML model using tabular data | 70 | Generating predictions using the recently trained model | 88 |

Deploying a model in Vertex AI    88    Generating predictions programmatically    93

Generating predictions    92    Summary    97

# 6

# Low-Code Options for Building ML Models    99

What is BQML?    99    Creating BQML models    114

Getting started with BigQuery    101    Hyperparameter tuning with BQML    122

Using BQML for feature    Evaluating trained models    126

transformations    102    Doing inference with BQML    129

Manual preprocessing    103    User exercise    131

Building ML models with BQML    113    Summary    131

# 7

# Training Fully Custom ML Models with Vertex AI    133

Technical requirements    133    Packaging a model to submit it to

Building a basic deep learning model    Vertex AI as a training job    147

with TensorFlow    134    Monitoring model training progress    156

Experiment – converting black-and-white    Evaluating trained models    160

images into color images    134    Summary    162

# 8

# ML Model Explainability    163

What is Explainable AI and why is it    Explainable AI features available in

important for MLOps practitioners?    164    Google Cloud Vertex AI    172

Building trust and confidence    164    Feature-based explanation techniques

available on Vertex AI    172

Explainable AI techniques    165    Using the model feature importance

Global versus local explainability    165    (SHAP-based) capability with AutoML

Techniques for image data    166    for tabular data    173

Techniques for tabular data    168    Exercise 1    173

Techniques for text data    170    Exercise 2    176

Example-based explanations                     186
Key steps to use example-based explanations    187
Exercise 3                                     187

Summary          188
References       188

# 9

# Model Optimizations – Hyperparameter Tuning and NAS                189

Technical requirements                     189
What is HPT and why is it
important?                                 190
What are hyperparameters?                  190
Why HPT?                                    190
Search algorithms                          190
Setting up HPT jobs on Vertex AI           191
What is NAS and how is it different
from HPT?                                  207

Search space                    208
Optimization method             208
Evaluation method               208
NAS on Vertex AI overview       209
NAS best practices              210
Summary                         211

# 10

# Vertex AI Deployment and Automation Tools – Orchestration through Managed Kubeflow Pipelines                213

Technical requirements                             214
Orchestrating ML workflows using
Vertex AI Pipelines (managed
Kubeflow pipelines)                                214
Developing Vertex AI Pipeline using Python         214
Pipeline components                                216
Orchestrating ML workflows using
Cloud Composer (managed Airflow)                   227
Creating a Cloud Composer environment              227
Vertex AI Pipelines versus Cloud
Composer                                           232

Getting predictions on Vertex AI       233
Getting online predictions             234
Getting batch predictions              237
Managing deployed models on
Vertex AI                              239
Multiple models – single endpoint      239
Single model – multiple endpoints      239
Compute resources and scaling          240
Summary                                240

## 11

# MLOps Governance with Vertex AI        241

**What is MLOps governance and
what are its key components?**        242

Data governance        242

Model governance        242

**Enterprise scenarios that
highlight the importance of
MLOps governance**        243

Scenario 1 – limiting bias in AI solutions        243

Scenario 2 – the need to constantly monitor
shifts in feature distributions        243

Scenario 3 – the need to monitor costs        244

Scenario 4 – monitoring how the training
data is sourced        244

**Tools in Vertex AI that can help
with governance**        245

Model Registry        245

Metadata Store        247

Feature Store        251

Vertex AI pipelines        253

Model Monitoring        254

Billing monitoring        257

**Summary**        258

**References**        258

# Part 3: Prebuilt/Turnkey ML Solutions Available in GCP

## 12

# Vertex AI – Generative AI Tools        261

**GenAI fundamentals**        262

GenAI versus traditional AI        262

Types of GenAI models        262

Challenges of GenAI        262

LLM evaluation        264

**GenAI with Vertex AI**        265

Understanding foundation models        265

What is a prompt?        271

Using Vertex AI GenAI models through
GenAI Studio        275

Example 1 – using GenAI Studio language
models to generate text        276

Example 2 – submitting examples along with
the text prompt in structured format to get
generated output in a specific format        277

Example 3 – generating images using GenAI
Studio (Vision)        280

Example 4 – generating code samples        281

**Building and deploying GenAI
applications with Vertex AI**        282

**Enhancing GenAI performance with
model tuning in Vertex AI**        286

Using Vertex AI supervised tuning        287

Safety filters for generated content        289

**Summary**        291

**References**        292

## 13

### Document AI – An End-to-End Solution for Processing Documents   293

Technical requirements                     294
What is Document AI?                       294
Document AI processors                     294

Overview of existing Document AI
processors                                 295

Using Document AI processors               296

Creating custom Document AI
processors                                 300
Summary                                    302

## 14

### ML APIs for Vision, NLP, and Speech                              303

Vision AI on Google Cloud                  304
Vision AI                                  304
Video AI                                   307

Translation AI on Google Cloud            308
Cloud Translation API                      309
AutoML Translation                         309
Translation Hub                            312

Natural Language AI on
Google Cloud                               313

AutoML for Text Analysis                   313
Natural Language API                       314
Healthcare Natural Language API            315

Speech AI on Google Cloud                  317
Speech-to-Text                             317
Text-to-Speech                             318

Summary                                    320

# Part 4: Building Real-World ML Solutions with Google Cloud

## 15

### Recommender Systems – Predict What Movies a User Would Like to Watch                                        323

Different types of recommender
systems                                    324
Real-world evaluation of recommender
systems                                    325

Deploying a movie recommender
system on Vertex AI                         327
Data preparation                           328
Model building                             330

Local model testing                                           331
Deploying the model on Google Cloud                           333
Using the model for inference                                 334

Summary                                                       338
References                                                    338

# 16

# Vision-Based Defect Detection System – Machines Can See Now!    339

Technical requirements                                        340
Vision-based defect detection                                 340
Dataset                                                       340
Importing useful libraries                                    341
Loading and verifying data                                    341
Checking few samples                                          341
Data preparation                                              343
Splitting data into train and test                            345
Final preparation of training and testing data               346
TF model architecture                                         347
Compiling the model                                           350
Training the model                                            350

Plotting the training progress                                351
Results                                                       352

Deploying a vision model to a Vertex
AI endpoint                                                   355
Saving model to Google Cloud Storage (GCS)                    355
Uploading the TF model to the Vertex Model
Registry                                                      356
Creating a Vertex AI endpoint                                 357
Deploying a model to the Vertex AI endpoint                   358

Getting online predictions from a
vision model                                                  359
Summary                                                       361

# 17

# Natural Language Models – Detecting Fake News Articles!    363

Technical requirements                                        363
Detecting fake news using NLP                                 364
Fake news classification with random forest                   364
About the dataset                                             364
Importing useful libraries                                    365
Reading and verifying the data                                365
NULL value check                                              366
Combining title and text into a single column                 366
Cleaning and pre-processing data                              366
Separating the data and labels                                368
Converting text into numeric data                             368
Splitting the data                                            369
Defining the random forest classifier                         369

Training the model                                            369
Predicting the test data                                      369
Checking the results/metrics on the
test dataset                                                  370
Confusion matrix                                              371

Launching model training on
Vertex AI                                                     372
Setting configurations                                        373
Initializing the Vertex AI SDK                                373
Defining the Vertex AI training job                           373
Running the Vertex AI job                                     374

BERT-based fake news classification    374
BERT for fake news classification                             375

Importing useful libraries                        375
The dataset                                       376
Data preparation                                  376
Splitting the data                                377
Creating data loader objects for batching         378
Loading the pre-trained BERT model                378

Scheduler                                         379
Training BERT                                     380
Loading model weights for evaluation              381
Calculating the accuracy of the test dataset      381
Classification report                             382

**Summary**                                       **383**

**Index**                                         **385**

**Other Books You May Enjoy**                     **398**

# Preface

Hello there! *The Definitive Guide to Google Vertex AI* is a comprehensive guide on accelerating the development and deployment of real-world ML solutions, with the help of the frameworks and best practices offered by Google as part of Vertex AI within Google Cloud.

Developing large-scale ML solutions and managing ML workflows in production is important for every business nowadays. Google has developed a unified data and AI platform, called Google Vertex AI, to help accelerate your ML journey and MLOps tools for workflow management.

This book is a complete guide that lets you explore all the features of Google Vertex AI, from an easy to advanced level, for end-to-end ML solution development. Starting from data management, model building, and experimentation to deployment, the Vertex AI platform provides you with tooling for no-code and low-code as well as fully customized approaches.

This book also provides a hands-on guide to developing and deploying some real-world applications on Google Cloud Platform, using technologies such as computer vision, NLP, and generative AI. Additionally, this book discusses some prebuilt/turnkey solution offerings from Google and shows you how to quickly integrate them into ML projects.

## Who this book is for

If you are a machine learning practitioner who wants to learn end-to-end ML solution development journey on Google Cloud Platform, using the MLOps best practices and tools offered by Google Vertex AI, this book is for you. Starting from data storage and data management, this book takes you through the Vertex AI offerings to build, experiment, optimize, and deploy ML solutions in a fast and scalable way. It also covers topics related to scaling, monitoring, and governing your ML workloads with the help of MLOps tooling on Google Cloud.

## What this book covers

*Chapter 1*, *Machine Learning Project Life Cycle and Challenges*, provides an introduction to a typical ML project's life cycle. It also highlights the common challenges and limitations of developing ML solutions for real-world use cases.

*Chapter 2*, *What Is MLOps, and Why Is It So Important for Every ML Team?* covers a set of practices usually known as MLOps that mature ML teams use as part of their ML development life cycle.

*Chapter 3, It's All about Data – Options to Store and Transform ML Datasets*, provides an overview of the different options available for storing data and analyzing data in Google Cloud. It also helps you to choose the best option based on your requirements.

*Chapter 4, Vertex AI Workbench – a One-Stop Tool for for AI/ML Development Needs*, demonstrates the use of a Vertex AI Workbench-based notebook environment for end-to-end ML solution development.

*Chapter 5, No-Code Options for Building ML Models*, covers GCP AutoML capabilities that can help users build state-of-the-art ML models, without the need for code or deep data science knowledge.

*Chapter 6, Low-Code Options for Building ML Models*, covers how to use **BigQuery ML (BQML)** to build and evaluate ML models using just SQL.

*Chapter 7, Training Fully Custom ML Models with Vertex AI*, explores how to develop fully customized ML solutions using the Vertex AI tooling available on Google Cloud. This chapter also shows you how to monitor training progress and evaluate ML models.

*Chapter 8, ML Model Explainability*, discusses concepts around ML model explainability and describes how to effectively incorporate explainable models into your ML solutions, using Vertex AI.

*Chapter 9, Model Optimizations – Hyperparameter Tuning and NAS*, explains the need for model optimization. It also covers two model optimization frameworks in detail – hyperparameter tuning and **Neural Architecture Search (NAS)**.

*Chapter 10, Vertex AI Deployment and Automation Tools – Orchestration through Managed Kubeflow Pipelines*, provides an overview of ML orchestrations and automation tools. This chapter further covers the implementation examples of ML workflow orchestration, using Cloud Composer and Vertex AI pipelines.

*Chapter 11, MLOps Governance with Vertex AI*, describes the different Google Cloud ML tools that can be used to deploy governance and monitoring controls.

*Chapter 12, Vertex AI – Generative AI Tools*, provides an overview of Vertex AI's recently launched generative AI features, such as Model Garden and Generative AI Studio.

*Chapter 13, Document AI – an End-to-End Solution for Processing Documents*, provides an overview of the document processing-related offerings on Google Cloud, such as OCR and Form Parser. This chapter also shows how to combine prebuilt and custom document processing solutions to develop a custom document processor.

*Chapter 14, ML APIs for Vision, NLP, and Speech*, provides an overview of the prebuilt state-of-the-art solutions from Google for computer vision, NLP, and speech-related use cases. It also shows you how to integrate them to solve real-world problems.

*Chapter 15, Recommender Systems – Predict What Movies a User Would Like to Watch*, provides an overview of popular approaches to building recommender systems and how to deploy one using Vertex AI.

*Chapter 16, Vision-Based Defect Detection System – Machines Can See Now*, shows you how to develop end-to-end computer vision-based custom solutions using Vertex AI tooling on Google Cloud, enabling you to solve real-world use cases.

*Chapter 17, Natural Language Models – Detecting Fake News Articles*, shows you how to develop NLP-related, end-to-end custom ML solutions on Google Cloud. This chapter explores a classical as well as a deep learning-based approach to solving the problem of detecting fake news articles.

## To get the most out of this book

You will need to have a basic understanding of machine learning and deep learning techniques. You also should have beginner-level experience with the Python programming language.

| Software/hardware for the coding exercises | Operating system requirements |
|---|---|
| Python 3.8 or later | Windows, macOS, or Linux |
| Google Cloud SDK | Windows, macOS, or Linux |
| A Google Cloud Platform account | N/A |

To ensure that you are using the correct Python library versions while executing the code samples, you can check out the GitHub repository of this book, where the code example notebooks also contain the version information.

If you are using the digital version of this book, we advise you to type the code yourself or access the code from the book's GitHub repository (a link is available in the next section). Doing so will help you avoid any potential errors related to the copying and pasting of code.

## Download the example code files

You can download the example code files for this book from GitHub at `https://github.com/PacktPublishing/The-Definitive-Guide-to-Google-Vertex-AI`. If there's an update to the code, it will be updated in the GitHub repository.

We also have other code bundles from our rich catalog of books and videos available at `https://github.com/PacktPublishing/`. Check them out!

## Conventions used

There are a number of text conventions used throughout this book.

`Code in text`: Indicates code words in text, database table names, folder names, filenames, file extensions, pathnames, dummy URLs, user input, and Twitter handles. Here is an example: "By default, the Jupyter server starts on port `8888`, but in case, this port is unavailable so it finds the next available port."

A block of code is set as follows:

```
export PROJECT=$(gcloud config list project --format     "value(core.
project)")

docker build . -f Dockerfile.example -t "gcr.io/${PROJECT}/tf-
custom:latest"

docker push "gcr.io/${PROJECT}/tf-custom:latest"
```

Any command-line input or output is written as follows:

```
$ mkdir css
$ cd css
```

**Bold**: Indicates a new term, an important word, or words that you see on screen. For instance, words in menus or dialog boxes appear in **bold**. Here is an example: "In the **Environment** field, select **Custom Container**."

> **Tips or important notes**
> Appear like this.

# Get in touch

Feedback from our readers is always welcome.

**General feedback**: If you have questions about any aspect of this book, email us at customercare@packtpub.com and mention the book title in the subject of your message.

**Errata**: Although we have taken every care to ensure the accuracy of our content, mistakes do happen. If you have found a mistake in this book, we would be grateful if you would report this to us. Please visit www.packtpub.com/support/errata and fill in the form.

**Piracy**: If you come across any illegal copies of our works in any form on the internet, we would be grateful if you would provide us with the location address or website name. Please contact us at copyright@packt.com with a link to the material.

**If you are interested in becoming an author**: If there is a topic that you have expertise in and you are interested in either writing or contributing to a book, please visit authors.packtpub.com.

## Share Your Thoughts

Once you've read *The Definitive Guide to Google Vertex AI*, we'd love to hear your thoughts! Scan the QR code below to go straight to the Amazon review page for this book and share your feedback.

https://packt.link/r/1-801-81526-7

Your review is important to us and the tech community and will help us make sure we're delivering excellent quality content.

# Download a free PDF copy of this book

Thanks for purchasing this book!

Do you like to read on the go but are unable to carry your print books everywhere?

Is your eBook purchase not compatible with the device of your choice?

Don't worry, now with every Packt book you get a DRM-free PDF version of that book at no cost.

Read anywhere, any place, on any device. Search, copy, and paste code from your favorite technical books directly into your application.

The perks don't stop there, you can get exclusive access to discounts, newsletters, and great free content in your inbox daily

Follow these simple steps to get the benefits:

1.  Scan the QR code or visit the link below

https://packt.link/free-ebook/978-1-80181-526-0

2.  Submit your proof of purchase
3.  That's it! We'll send your free PDF and other benefits to your email directly

# Part 1:
# The Importance of MLOps in a Real-World ML Deployment

In this part, you will get an overview of the life cycle of a typical real-world **machine learning (ML)** project. You will also learn about common challenges encountered during the development of ML applications and some key limitations of an ML framework. Finally, you will learn about the **machine learning operations (MLOps)** practice and its importance in ML deployments.

This part has the following chapters:

- *Chapter 1, Machine Learning Project Life Cycle and Challenges*
- *Chapter 2, What Is MLOps, and Why Is It So Important for Every ML Team?*

# 1
# Machine Learning Project Life Cycle and Challenges

Today, **machine learning (ML)** and **artificial intelligence (AI)** are integral parts of business strategy for many organizations, and more organizations are using them every year. The major reason for this adoption is the power of ML and AI solutions to garner more revenue, brand value, and cost savings. This increase in the adoption of AI and ML demands more skilled data and ML specialists and technical leaders. If you are an ML practitioner or beginner, this book will help you become a confident ML engineer or data scientist with knowledge of Google's best practices. In this chapter, we will discuss the basics of the life cycle and the challenges and limitations of ML when developing real-world applications.

ML projects often involve a defined set of steps from problem statements to deployments. It is essential to understand the importance and common challenges involved with these steps to complete a successful and impactful project. In this chapter, we will discuss the importance of understanding the business problem, the common steps involved in a typical ML project life cycle, and the challenges and limitations of ML in detail. This will help new ML practitioners understand the basic project flow; plus, it will help create a foundation for forthcoming chapters in this book.

This chapter covers the following topics:

- ML project life cycle
- Common challenges in developing real-world ML solutions
- Limitations of ML

## ML project life cycle

In this section, we will learn about the typical life cycle of an ML project, from defining the problem to model development, and finally, to the operationalization of the model. *Figure 1.1* shows the high-level steps almost every ML project goes through. Let's go through all these steps in detail.

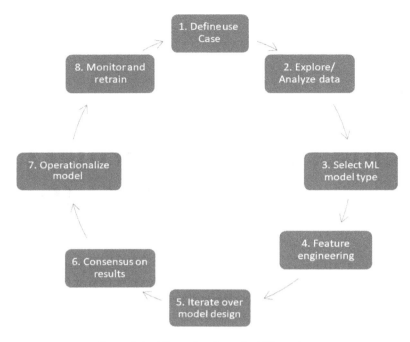

Figure 1.1 – Life cycle of a typical ML project

Just like the **Software Development Life Cycle (SDLC)**, the **Machine Learning Project/Development Lifecycle (MDLC)** guides the end-to-end process of ML model development and operationalization. At a high level, the life cycle of a typical ML project in an enterprise setting remains somewhat consistent and includes eight key steps:

1.  Define the ML use case: The first step of any ML project is where the ML team works with business stakeholders to assess the business needs around predictive analytics and identifies a use case where ML can be used, along with some success criteria, performance metrics, and possible datasets that can be used to build the models.

    For example, if the sales/marketing department of an insurance company called ABC Insurance Inc. wants to better utilize its resources to target customers who are more likely to buy a certain product, they might approach the ML team to build a solution that can sift through all possible leads/customers and, based on the data points for each lead (age, prior purchase, length of policy history, income level, etc.), identify the customers who are most likely to buy a policy. Then the sales team can ask their customer representatives to prioritize reaching out to these customers instead of calling all possible customers blindly. This can significantly improve the outcome of outbound calls by the reps and improve the sales-related KPIs.

Once the use case is defined, the next step is to define a set of KPIs to measure the success of the solution. For this sales use case, this could be the customer sign-up rate—what percentage of the customers whom sales reps talk to sign up for a new insurance policy?

To measure the effectiveness of the ML solution, the sales team and the ML team might agree to measure the increase or decrease in customer sign-up rate once the ML model is live and iteratively improve on the model to optimize the sign-up rate.

At this stage, there will also be a discussion about the possible datasets that can be utilized for the model training. These could include the following:

- Internal customer/product datasets being generated by marketing and sales teams, for example, customer metadata, such as their age, education profile, income level, prior purchase behavior, number and type of vehicles they own, etc.

- External datasets that can be acquired through third parties; for example, an external marketing consultancy might have collected data about the insurance purchase behavior of customers based on the car brand they own. This additional data can be used to predict how likely they are to purchase the insurance policy being sold by ABC Insurance Inc.

2.  Explore/analyze data: The next step is to do a detailed analysis of the datasets. This is usually an iterative process in which the ML team works closely with the data and business SMEs to better understand the nuances of the available datasets, including the following:

- Data sources

- Data granularity

- Update frequency

- Description of individual data points and their business meaning

This is a key step where data scientists/ML engineers analyze the available data and decide what datasets might be relevant to the ML solution being considered, analyze the robustness of the data, and identify any gaps. Issues that the team might identify at this stage could relate to the cleanliness and completeness of data or problems with the timely availability of the data in production. For example, the age of the customer could be a great indicator of their purchase behavior, but if it's an optional field in the customer profile, only a handful of customers might have provided their date of birth or age.

So, the team would need to figure out if they want to use the field and, if so, how to handle the samples where age is missing. They could also work with sales and marketing teams to make the field a *required field* whenever a new customer requests an insurance quote online and generates a lead in the system.

3.  Select ML model type: Once the use case has been identified along with the datasets that can possibly be used to train the model, the next step is to consider the types of models that can be used to achieve the requirements. We won't go too deep into the topic of general model

selection here since entire books could be written on the topic, but in the next few chapters, you will see what different model types can be built for specific use cases in Vertex AI. At a very high level, the key considerations at this stage are as follows:

- Type of model: For example, for the insurance customer/lead ranking example, we could build a classification model that will predict whether a new customer is *high/medium/low* in terms of their likelihood to purchase a policy. Or a regression model could be built to output a sales probability number for each likely customer.

- Does the conventional ML model satisfy our requirements or do we need a deep learning model?

- Explainability requirements: Does the use case require an explanation for each prediction as to why the sample was classified a certain way?

- Single versus ensemble model: Do we need a single model to give us the final prediction, or do we need to employ a set of interconnected models that feed into each other? For example, a first model might assign a customer to a particular customer group, and the next model might use that grouping to identify the final likelihood of purchase.

- Separation of models: For example, sometimes we might build a single global model for the entire customer base, or we might need separate models for each region due to significant differences in products and user behavior in different regions.

4. Feature engineering: This process is usually the most time-consuming and involves several steps:

   I. Data cleanup–Imputing missing values where possible, dropping fields with too many missing values

   II. Data and feature augmentation–Joining datasets to bring in additional fields, and cross-joining existing features to generate new features

   III. Feature analysis–Calculating feature correlation and analyzing collinearity, checking for data leakage in features

Again, since this is an extremely broad topic, we are not diving too deep into it and suggest you refer to other books on this topic.

5. Iterate over the model design/build: The actual design and build of the ML model is an iterative process involving the following key steps:

   I. Select model architecture

   II. Split acquired data into train/validation/test subsets

   III. Run model training experiments, tune hyperparameters

   IV. Evaluate trained models with the test dataset

   V. Rank and select the best models

*Figure 1.2* shows the typical ML model development life cycle:

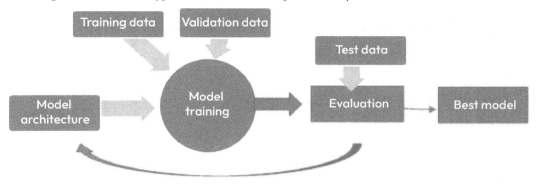

Figure 1.2 – ML model development life cycle

6.  Consensus on results: Once a satisfactory model has been obtained, the ML team shares the results with the business stakeholders to ensure the results fully align with the business needs and performs additional optimizations and post-processing steps to make the model predictions usable by the business. To assure business stakeholders that ML solution is aligned with the business goals and is accurate enough to drive value, ML teams could use one of a number of approaches:

    - Evaluate using historical test datasets: ML teams can run historical data through the new ML models and evaluate the predictions against the ground truth values. For example, in the insurance use case discussed previously, the ML team can take last month's data on customer leads and use the ML model to predict which customers are most likely to purchase a new insurance policy. Then they can compare the model's predictions against the actual purchase history from the previous month and see how accurate the model's predictions were. If the model's output is close to the real purchase behavior of customers, then the model is working as desired, and this information can be presented to business stakeholders to convince them of the ML solution's efficacy in driving additional revenue. On the contrary, if the model's output significantly deviates from the customer's behavior, the ML team needs to go back and work on improving the model. This usually is an iterative process and can take a number of iterations, depending on the complexity of the model.

    - Evaluate with live data: In some scenarios, an organization might decide to conduct a small pilot in a production environment with real-time data to assess the performance of the new ML model. This is usually done in the following scenarios:

        - When there is no historical data available to conduct the evaluation or where testing with historical data is not expected to be an accurate; for example, during the onset of COVID, customer behavior patterns abruptly changed to the extent that testing with any historical data became nearly useless

* When there is an existing model in production being used for critical real-time predictions, the sanity check for the new model needs to be performed not just in terms of its accuracy but also its subtle impact on downstream KPIs such as revenue per user session

In such cases, teams might deploy the model in production, divert a small number of prediction requests to the newer model, and periodically compare the overall impact on the KPIs. For example, in the case of a recommendation model deployed on an e-commerce website, a recommendation model might start recommending products that are comparatively cheaper than the predictions from the older model already live in production. In this scenario, the likelihood of a customer completing a purchase would go up, but at the same time, the revenue generated per user session would decrease, impacting overall revenue for the organization. So, although it might seem like the ML model is working as designed, it might not be considered a success by the business/sales stakeholders, and more discussions would be required to optimize it.

7. Operationalize model: Once the model has been approved for deployment in production, the ML team will work with their organization's IT and data engineering teams to deploy the model so that other applications can start utilizing it to generate insights. Depending on the size of the organization, there can be significant overlap in the roles these teams play.

The actual deployment architecture would depend on the following:

* Prediction SLAs – Ranging from periodic batch jobs to solutions that require sub-second prediction performance.

* Compliance requirements – Can the user data be sent to third-party cloud providers, or does it need to always reside within an organization's data centers?

* Infrastructure requirements – This depends on the size of the model and its compute requirements. Small models can be served from a shared compute node. Some large models might need a large GPU-connected node.

We will discuss this topic in detail in later chapters, but the following figure shows some key components you might consider as part of your deployment architecture.

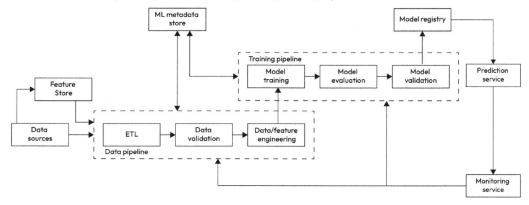

Figure 1.3 – Key components of ML model training and deployment

8.  Monitor and retrain: It might seem as if the ML team's job is done once the model has been operationalized, but in real-world deployments, most models require periodic or sometimes constant monitoring to ensure the model is operating within the required performance thresholds. Model performance could become sub-optimal for several reasons:

    *   Data drift: Changes in data being used to generate predictions could change significantly and impact the model's performance. As we discussed before, during COVID, customer behavior changed significantly. Models that were trained on pre-COVID customer behavior data were not equipped to handle this sudden change in usage patterns. Change due to the pandemic was relatively rare but high-impact, but there are plenty of other smaller changes in prediction input data that might impact your model's performance adversely. The impact could range from a subtle drop in accuracy to a model generating erroneous responses. So, it is important to keep an eye on the key performance metrics of your ML solution.

    *   Change prediction request volume: If your solution was designed to handle 100 requests per second but now is seeing periodic bursts in traffic of around 1,000 requests per second, your solution might not be able to keep up with the demand, or latency might go above acceptable levels. So, your solution also needs to have monitoring and certain levels of auto-scaling built in to handle such scenarios. For larger changes in traffic volumes, you might even need to completely rethink the serving architecture.

    There would be scenarios where through monitoring, you will discover that your ML model no longer meets the prediction accuracy and requires retraining. If the change in data patterns is expected, the ML team should design the solution to support automatic periodic retraining. For example, in the retail industry, product catalogs, pricing, and promotions constantly evolve, requiring regular retraining of the models. In other scenarios, the change might be gradual or unexpected, and when the monitoring system alerts the ML team of the model performance degradation, they need to take a call on retraining the model with more recent data, or maybe even completely rebuilding the model with new features.

Now that we have a good idea of the life cycle of an ML project, let's learn about some of the common challenges faced by ML developers when creating and deploying ML solutions.

# Common challenges in developing real-world ML solutions

A real-world ML project is always filled with some unexpected challenges that we get to experience at different stages. The main reason for this is that the data present in the real world, and the ML algorithms, are not perfect. Though these challenges hamper the performance of the overall ML setup, they don't prevent us from creating a valuable ML application. In a new ML project, it is difficult to know the challenges up front. They are often found during different stages of the project. Some of these challenges are not obvious and require skilled or experienced ML practitioners (or data scientists) to identify them and apply countermeasures to reduce their effect.

In this section, we will understand some of the common challenges encountered during the development of a typical ML solution. The following list shows some common challenges we will discuss in more detail:

- Data collection and security
- Non-representative training set
- Poor quality of data
- Underfitting of the training dataset
- Overfitting of the training dataset
- Infrastructure requirements

Now, let's learn about each of these common challenges in detail.

## Data collection and security

One of the most common challenges that organizations face is data availability. ML algorithms require a large amount of good-quality data in order to provide quality results. Thus, the availability of raw data is critical for a business if it wants to implement ML. Sometimes, even if the raw data is available, gathering data is not the only concern; we often need to transform or process the data in a way that our ML algorithm supports.

Data security is another important challenge that is very frequently faced by ML developers. When we get data from a company, it is essential to differentiate between sensitive and non-sensitive information to implement ML correctly and efficiently. The sensitive part of data needs to be stored in fully secured servers (storage systems) and should always be kept encrypted. Sensitive data should be avoided for security purposes, and only the less-sensitive data access should be given to trusted team members working on the project. If the data contains **Personally Identifiable Information** (**PII**), it can still be used by anonymizing it properly.

## Non-representative training data

A good ML model is one that performs equally well on unseen data and training data. It is only possible when your training data is a good representative of most possible business scenarios. Sometimes, when the dataset is small, it may not be a true representative of the inherent distribution, and the resulting model may provide inaccurate predictions on unseen datasets despite having high-quality results on the training dataset. This kind of non-representative data is either the result of sampling bias or the unavailability of data. Thus, an ML model trained on such a non-representative dataset may have less value when it is deployed in production.

If it is impossible to get a true representative training dataset for a business problem, then it's better to limit the scope of the problem to only the scenarios for which we have a sufficient amount of training samples. In this way, we will only get known scenarios in the unseen dataset, and the model should provide quality predictions. Sometimes, the data related to a business problem keeps changing with time, and it may not be possible to develop a single static model that works well; in such cases, continuous retraining of the model on the latest data becomes essential.

## Poor quality of data

The performance of ML algorithms is very sensitive to the quality of training samples. A small number of outliers, missing data cases, or some abnormal scenarios can affect the quality of the model significantly. So, it is important to treat such scenarios carefully while analyzing the data before training any ML algorithm. There are multiple methods for identifying and treating outliers; the best method depends upon the nature of the problem and the data itself. Similarly, there are multiple ways of treating the missing values as well. For example, mean, median, mode, and so on are some frequently used methods to fill in missing data. If the training data size is sufficiently large, dropping a small number of rows with missing values is also a good option.

As discussed, the quality of the training dataset is important if we want our ML system to learn accurately and provide quality results on the unseen dataset. It means that the data pre-processing part of the ML life cycle should be taken very seriously.

## Underfitting the training dataset

Underfitting an ML model means that the model is too simple to learn the inherent information or structure of the training dataset. It may occur when we try to fit a non-linear distribution using a linear ML algorithm such as linear regression. Underfitting may also occur when we utilize only a minimal set of features (that may not have much information about the target distribution) while training the model. This type of model can be too simple to learn the target distribution. An underfitted model learns too little from the training data and, thus, makes mistakes on unseen or test datasets.

There are multiple ways to tackle the problem of underfitting. Here is a list of some common methods:

- Feature engineering – add more features that represent target distribution
- Non-linear algorithms – switch to a non-linear algorithm if the target distribution is not linear
- Removing noise from the data
- Add more power to the model – increase trainable parameters, increase depth or number of trees in tree-based ensembles

Just like underfitting the model on training data, overfitting is also a big issue. Let's deep dive into it.

## Overfitting the training dataset

The overfitting problem is the opposite of the underfitting problem. Overfitting is the scenario when the ML model learns too much unnecessary information from the training data and fails to generalize on a test or unseen dataset. In this case, the model performs extremely well on the training dataset, but the metric value (such as accuracy) is very low on the test set. Overfitting usually occurs when we implement a very complex algorithm on simple datasets.

Some common methods to address the problem of overfitting are as follows:

- Increase training data size – ML models often overfit on small datasets
- Use simpler models – When problems are simple or linear in nature, choose simple ML algorithms
- Regularization – There are multiple regularization methods that prevent complex models from overfitting on the training dataset
- Reduce model complexity – Use a smaller number of trainable parameters, train for a smaller number of epochs, and reduce the depth of tree-based models

Overfitting and underfitting are common challenges and should be addressed carefully, as discussed earlier. Now, let's discuss some infrastructure-related challenges.

## Infrastructure requirements

ML is expensive. A typical ML project often involves crunching large datasets with millions or billions of samples. Slicing and dicing such datasets requires a lot of memory and high-end multi-core processors. Additionally, once the development of the project is complete, dedicated servers are required to deploy the models and match the scale of consumers. Thus, business organizations willing to practice ML need some dedicated infrastructure to implement and consume ML efficiently. This requirement increases further when working with large, deep learning models such as transformers, **large language models** (LLMs), and so on. Such models usually require a set of accelerators, **graphical processing units** (GPUs), or **tensor processing units** (TPUs) for training, finetuning, and deployment.

As we have discussed, infrastructure is critical for practicing ML. Companies that lack such infrastructure can consult with other firms or adopt cloud-based offerings to start developing ML-based applications.

Now that we understand the common challenges faced during the development of an ML project, we should be able to make more informed decisions about them. Next, let's learn about some of the limitations of ML.

# Limitations of ML

ML is very powerful, but it's not the answer to every single problem. There are problems that ML is just not suitable for, and there are some cases where ML can't be applied due to technical or business constraints. As an ML practitioner, it is important to develop the ability to find relevant business problems where ML can provide significant value instead of applying it blindly everywhere. Additionally, there are algorithm-specific limitations that can render an ML solution not applicable in some business applications. In this section, we will learn about some common limitations of ML that should be kept in mind while finding relevant use cases.

Keep in mind that the limitations we are discussing in this section are very general. In real-world applications, there are more limitations possible due to the nature of the problem we are solving. Some common limitations that we will discuss in detail are as follows:

- Data-related concerns
- Deterministic nature of problems
- Lack of interpretability and reproducibility
- Concerns related to cost and customizations
- Ethical concerns and bias

Let's now deep dive into each of these common limitations.

## Data-related concerns

The quality of an ML model highly depends upon the quality of the training data it is provided with. Data present in the real world is often noisy, incomplete, unlabeled, and sometimes unusable. Moreover, most supervised learning algorithms require large amounts of properly labeled training data to produce good results. The training data requirements of some algorithms (e.g., deep learning) are so high that even manually labeling data is not an option. And even if we manage to label the data manually, it is often error-prone due to human bias.

Another major issue is incompleteness or missing data. For example, consider the problem of automatic speech recognition. In this case, model results are highly biased toward the accent present in the training dataset. A model that is trained on the American accent doesn't produce good results on other accented speech. Since accents change significantly as we travel to different parts of the world, it is hard to gather and label relevant amounts of training data for every possible accent. For this reason, developing a single speech recognition model that works for everyone is not yet feasible, and thus, the tech giants providing speech recognition solutions often develop accent-specific models. Developing a new model for each new accent is not very scalable.

## Deterministic nature of problems

ML has achieved great success in solving some highly complex problems, such as numerical weather prediction. One problem with most of the current ML algorithms is that they are stochastic in nature and thus cannot be trusted blindly when the problem is deterministic. Considering the case of numerical weather prediction, today we have ML models that can predict rain, wind speed, air pressure, and so on, with acceptable accuracy, but they completely fail to understand the physics behind real weather systems. For example, an ML model might provide negative value estimations of parameters such as density.

However, it is very likely that these kinds of limitations can be overcome in the near future. Future research in the field of ML might discover new algorithms that are smart enough to understand the physics of our world. Such models will open infinite possibilities in the future.

## Lack of interpretability and reproducibility

One major issue with many ML algorithms (and often with neural networks) is the lack of interpretability of results. Many business applications, such as fraud detection and disease prediction, require a justification for model results. If an ML model classifies a financial transaction as fraud, it should also provide solid evidence for the decision; otherwise, this output may not be useful for the business. Deep learning or neural network models often lack interpretability, and the explainability of such models is an active area of research. Multiple methods have been developed for model interpretability or explainability purposes. Though these methods can provide some insights into the results, they are still far from the actual requirements.

Reproducibility, on the other hand, is another complex and growing issue with ML solutions. Some of the latest research papers might show us great improvements in results using some technological advancements on a fixed set of datasets, but the same method may not work in real-world scenarios. Secondly, ML models are often unstable, which means that they produce different results when trained on different partitions of the dataset. This is a challenging situation because models developed for one business segment may be completely useless for another business segment, even though the underlying problem statement is similar. This makes them less reusable.

## Concerns related to cost and customizations

Developing and maintaining ML solutions is often expensive, more so in the case of deep learning algorithms. Development costs may come from employing highly skilled developers as well as the infrastructure needed for data analytics and ML experimentation. Deep learning models usually require high-compute resources such as GPUs and TPUs for training and experimentation. Running a hyperparameter tuning job with such models is even more costly and time-consuming. Once the model is ready for production, it requires dedicated resources for deployment, monitoring, and maintenance. This cost further increases as you scale your deployments to serve a large number of customers, and even more if there are very low latency concerns. Thus, it is very important to understand the value that our solution is going to bring before jumping into the development phase and check whether it is worth the investment.

Another concern with the ML solutions is their lack of customizations. ML models are often very difficult to customize, meaning it can be hard to change their parameters or make them adapt to new datasets. Pre-built general-purpose ML solutions often do not work well on specific business use cases, and this leaves them with two choices – either to develop the solution from scratch or customize the prebuilt general-purpose solutions. Though the customization of prebuilt models seems like a better choice here, even the customization is not easy in the case of ML models. ML model customization requires a skilled set of data engineers and ML specialists with a deep understanding of technical concepts such as deep learning, predictive modeling, and transfer learning.

## Ethical concerns and bias

ML is quite powerful and is adopted today by many organizations to guide their business strategy and decisions. As we know, some of these ML algorithms are *black boxes*; they may not provide reasons behind their decisions. ML systems are trained on a finite set of datasets, and they may not apply to some real-world scenarios; if those scenarios are encountered in the future, we can't tell what decision the ML system will take. There might be ethical concerns related to such black-box decisions. For example, if a self-driving car is involved in a road accident, whom should you blame – the driver, the team that developed the AI system, or the car manufacturer? Thus, it is clear that the current advancements in ML and AI are not suitable for ethical or moral decision-making. Also, we need a framework to solve ethical concerns involving ML and AI systems.

The accuracy and speed of ML solutions are often commendable, but these solutions cannot always be trusted to be fair and unbiased. Consider AI software that recognizes faces or objects in a given image; this system could go wrong on photos where the camera is not able to capture racial sensitivity properly, or it may classify a certain type of dog (that is somewhat similar to a cat) as a cat. This kind of bias may come from a biased set of training or testing datasets used for developing AI systems. Data present in the real world is often collected and labeled by humans; thus, the bias that exists in humans is transferred into AI systems. Avoiding bias completely is impossible as we all are humans and are thus biased, but there are measures that can be taken to reduce it. Establishing a culture of ethics and building teams from diverse backgrounds can be a good step to reduce bias to a certain extent.

# Summary

ML is an integral part of any business strategy and decisions for many organizations today, thus it is very important to do it right. In this chapter, we learned about the general steps involved in a typical ML project development life cycle and their significance. We also highlighted some common challenges that ML practitioners face while undergoing project development. Finally, we listed some of the common limitations of ML in real-world scenarios to help us choose the right business problem and a fitting ML algorithm to solve it.

In this chapter, we learned about the importance of choosing the right business problem in order to deliver the maximum impact using ML. We also learned about the general flow of a typical ML project. We should now be confident about identifying the underlying ML-related challenges in a business process and making informed decisions about them. Finally, we have learned about the common limitations of ML algorithms, and it will help us apply ML in a better way to get the best out of it.

Just developing a high-performing ML model is not enough. The real value comes when it is deployed and used in real-world applications. Taking an ML model to production is not trivial and should be done in the right way. The next chapter is all about the guidelines and best practices to follow while operationalizing an ML model and it is going to be extremely important to understand it thoroughly before jumping into the later chapters of this book.

# 2

# What Is MLOps, and Why Is It So Important for Every ML Team?

**Machine learning operations** (**MLOps**) is a pivotal practice for modern ML teams, encompassing the blend of technological and operational best practices. At its heart, MLOps seeks to address the challenges of productionizing ML models and fostering better collaboration between data scientists and IT teams. With the rapid advancements in technology and increasing reliance on ML solutions, MLOps is becoming the backbone of a sustainable and scalable ML strategy. This chapter will delve deep into the essence of MLOps, detailing its significance, its various maturity levels, and the role of Google's Vertex AI in facilitating MLOps. By the end of this chapter, you will be equipped with a robust understanding of MLOps principles and what tools in Vertex AI can be used to implement those principles.

In this chapter, we will cover the following topics:

- Why is MLOps important?
- MLOps maturity levels
- How can Vertex AI help with implementing ML Ops?

Let's embark on this enlightening journey to master MLOps on Vertex AI.

## Why is MLOps important?

As the development and integration of ML models become more and more common in today's world, the need for a robust operational framework has become more critical than ever. MLOps aims to address this requirement by streamlining the entire process of developing, deploying, and monitoring ML models. In this section, we will discuss the importance of MLOps due to the following aspects:

- **Standardizing and automating ML workflows**

  MLOps aims to standardize and automate various stages of the ML life cycle, from data ingestion and preprocessing to model training, evaluation, and deployment. By doing so, it minimizes the likelihood of human errors, facilitates reproducibility, and improves overall efficiency. Google's Vertex AI offers managed services for each stage of the ML workflow, which helps organizations achieve consistency, automate processes, and reduce operational overhead.

- **Monitoring and managing model performance**

  One of the key aspects of MLOps is continuously monitoring and managing the performance of deployed models. This is crucial, as the effectiveness of ML models may degrade over time due to changes in data distribution, unforeseen edge cases, or evolving user behaviors. Google's Vertex AI supports MLOps by providing tools for monitoring model performance and generating alerts when performance thresholds are breached. Furthermore, it enables seamless integration with other monitoring and logging services in the Google Cloud ecosystem.

- **Ensuring scalability and flexibility**

  MLOps facilitates the scaling of ML solutions by providing a framework that can easily accommodate increased data volumes, more complex models, and additional infrastructure requirements. Google's Vertex AI is built to handle these demands, offering a range of services and tools that scale automatically, support distributed training and prediction, and allow users to choose the optimal hardware configurations for their specific use cases.

- **Security and compliance**

  MLOps emphasizes the importance of security and compliance in ML workflows, ensuring that data privacy and regulatory requirements are met. Google's Vertex AI supports these objectives by providing a secure environment for model training, storage, and deployment. Its integration with Google Cloud's **Identity and Access Management** (**IAM**) enables fine-grained control over access to resources, while encryption options protect data both at rest and in transit.

While developing ML models, there can be many shortcuts taken or complexities overlooked that accumulate "debt" over time, leading to system maintenance difficulties or failures in the future. As an ML solution deployed in production evolves, developers might add more features to improve accuracy and add overall value for users. Each feature often has its own specific way of preprocessing, normalizing, and configuring parameters. Over time, managing these configurations and related changes can become increasingly complex and the solutions might accumulate significant technical debt. MLOps provides the necessary practices and tooling to manage and mitigate these debts, ensuring sustainable and scalable ML deployments.

In summary, MLOps plays a vital role in streamlining ML workflows, enhancing collaboration, and ensuring that models are secure, scalable, and well maintained. Google's Vertex AI, with its comprehensive suite of tools and services, empowers organizations to embrace MLOps and unlock the full potential of their ML solutions.

Now, let's look at the different levels of maturity or complexity of typical MLOps implementations seen in the industry. Keep in mind that the architectures we will discuss are just representative samples of different levels of complexity. In the real world, you will see many different variations of these implementations, and each implementation is as unique as the organization using it.

# Implementing different MLOps maturity levels

Most new ML teams and organizations go through a phased MLOps journey as they build and refine their MLOps strategy. They usually start with a fully manual step-by-step process where data science/data engineering teams take an extremely manual, ad hoc approach to building and deploying models. Once a few models have been deployed and stabilized in production, it slowly becomes apparent that this manual process is not very scalable and that the team needs to put some processes and automation in place.

At this point, as issues arise in production, it also becomes apparent that this ad hoc approach is not easily auditable or reproducible. As the usage of the ML solution grows, it graduates from being just an experiment to something the organization becomes increasingly dependent on. Compliance teams and leadership also start making requests to make the model deployment process more well organized and auditable to ensure compliance with the company's IT policies. At this stage, the team leadership will need to put together a high-level MLOps strategy and implementation roadmap based on the resources they have at their disposal and the solution roadmap of projects they are working on. Some organizations decide to spend their time building the entire MLOps stack upfront, while others might decide to step through the different maturity levels one step at a time.

Now, let's look at the different maturity levels most ML organizations typically progress through.

## MLOps maturity level 0

This is the phase where an organization has just started experimenting with ML solutions and has not prepared a well-baked MLOps strategy in terms of what the standard process and tooling would look like as they scale up their ML adoptions. At this stage, the landscape looks like the following:

- The organization has just 1-2 models deployed in production for each business unit
- AI/ML development is handled by a small centralized team of data scientists
- The focus is on deployment speed instead of consistent processes
- The choice of ML tools is unclear, and the leadership wants to figure out what works and what doesn't before committing to a particular ML platform
- Most, if not all, steps are manual

The following diagram shows different key components of an MLOps solution at maturity level 0:

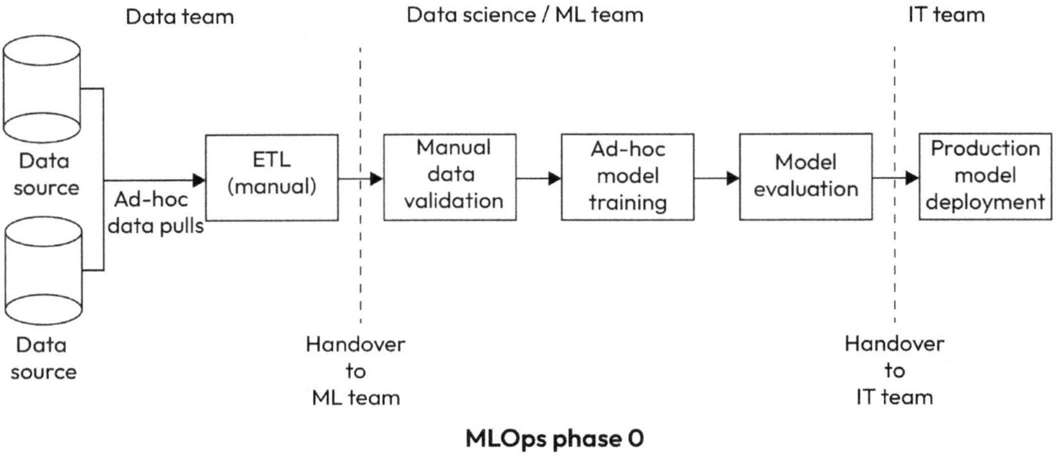

Figure 2.1 – MLOps solution: maturity level 0

As you can see in the preceding diagram, at MLOps maturity level 0, most of the handovers from one process to the next are manual. This ensures a short path to production deployment (when needed) by not requiring a build and test of extensive automation processes. The downside is obviously the significant time the team needs to spend every time the pipeline needs to be run. So, the ML engineering teams need to have a roadmap/strategy in place to add automation over time.

## MLOps maturity level 1 – automating basic ML steps

A key characteristic of maturity level 1 in MLOps is fully automated data and ML model training pipelines. At this stage, full data acquisition and model retraining can be triggered with a single click of a button or an API call.

One of the most significant bottlenecks that are most apparent to the team closest to the solution development is the data acquisition and model training process. In the *MLOps maturity level 0* section, most of the steps around data ingestion and model builds were done manually and are prone to errors even when repeating similar steps with slightly modified parameters. So, the data and ML model training pipelines end up being the very first components to get automated. It also helps that most of the data and ML pipelining work is being done within the same teams; that is, data pipeline work is done by the data team and the ML model design and building is done by the data science/ML team. This reduces the cross-team dependency on automation. There is obviously still a dependence on the IT team to ensure the availability of the right orchestration and automation tools.

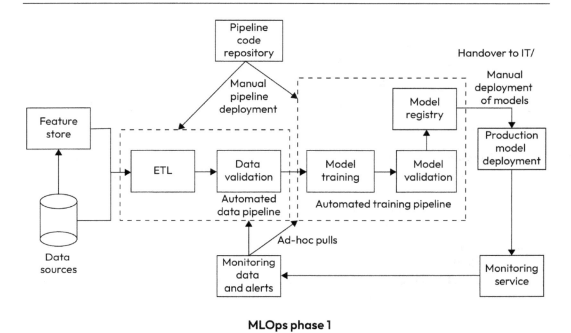

**MLOps phase 1**

Figure 2.2 – MLOps maturity level 1

As shown in the preceding diagram, most components in the MLOps pipeline are automated at this maturity level.

The components of the ML pipeline that are automated are the following:

- **Data imports**: The process of importing the latest data is almost entirely automated and started through external triggers that indicate the availability of newer data.

- **Feature/data engineering**: Automatically triggered as soon as preceding data transfer steps are completed.

- **Data validation**: At this stage, data validation is primarily handled through the orchestrated scripts triggered at the end of data/feature engineering steps.

- **Model training**: Once the data validation is successfully completed, the orchestration layer triggers the model training step, to rebuild the model with newly available data.

- **Model validation**: Automated benchmarking of the model's predictive performance, done through orchestrated scripts to ensure the model meets the business and technical requirements before being approved for production deployment.

- **Registration of new model in the centralized model registry**: Once the model has been validated, it gets pushed to a centralized model registry that keeps a catalog of all approved models.

The other new components that get introduced at this stage to support the automation are the following:

- **Code Repository**: To ensure consistency in the data and pipeline executions, creating a centralized repository to host the pipeline code/logic is important. For this, enterprises use their in-house Git implementations, tools such as GitHub or GitLab, or cloud-based code repository tools such as **Google Cloud Platform** (**GCP**) Cloud Source Repositories. (It's beyond the scope of this book to cover these tools in detail.)

- **Feature Store**: Centralized catalog/repository of features that can make experiment automation/ reproducibility easier and more consistent. This can be done using standalone specialized tools, such as **FEAST** or Vertex AI Feature Store, or a custom implementation using standard data warehouse tools. Vertex AI Feature Store is discussed in detail in *Chapter 11*, *MLOps Governance with Vertex AI*.

With these fundamental MLOps components in place, your solution would be a lot more efficient and a lot less prone to human errors.

## MLOps maturity level 2 – automated model deployments

Key characteristics of MLOps maturity level 2 are continuous deployment capabilities, which can automatically deploy any new ML models in the production environment, and triggering the creation of new ML models based on the triggers from monitoring services when shifts are detected in model accuracy.

The following diagram depicts a representative MLOps implementation with maturity level 2.

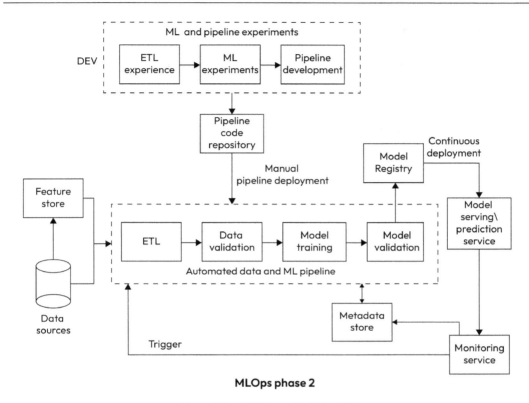

**MLOps phase 2**

Figure 2.3 – MLOps maturity level 2

Once the data acquisition and ML modeling portion have been automated in the maturity level 2 architecture, the next bottleneck that becomes apparent is the integration of the model build side with the model deployment side. This has two components:

- The process of taking a newly trained model that has passed the validation step and deploying it in production. For example, a new model that passes the validation tests should automatically get deployed in production.

- Taking the data collected by the monitoring service, turning it into actionable insights, and triggering updates to models. For example, a drift in incoming feature values in production can trigger the retraining of the ML model with newer data.

The other new component that gets introduced at this stage to support the automation is Metadata Store. This is the repository of all metadata that gets generated during every step of data acquisition, feature generation, model development, and deployment. This becomes important during automated deployments and retraining as the source of the model development history. For example, to monitor the model in production for data drift or training skew, the monitoring process needs access to the model's history to get the dataset on which it was trained. The other use case not related to automation is that Metadata Store can enable model auditing. If during the model development process all key parameters around the data and model are being logged in to Metadata Store, then it can provide end-to-end visibility around the model to the model auditors. An example of this is shown in the following figure, where you can see the journey of the model:

Figure 2.4 – Vertex AI Metadata Store

Now let us see how you can use Vertex AI to implement end-to-end MLOps solutions.

## How can Vertex AI help with implementing MLOps?

Google Cloud Vertex AI is a platform that provides tools and resources for the end-to-end implementation of the ML development life cycle.

Vertex AI can help with MLOps by providing features such as automated model building and deployment, model versioning and tracking, and monitoring and managing models in production. Additionally, it provides tools for collaboration and shared access to resources, allowing teams to work together on large and distributed ML projects.

Vertex AI can also help with other aspects of MLOps, such as the following:

- **Data management**: Vertex AI can help with data preparation, labeling, and management, which are crucial for building accurate ML models

- **Experimentation**: Vertex AI can help track and manage experiments, including comparing and selecting the best models

- **Model governance**: Vertex AI can help manage model access and permissions and monitor models for drift and compliance

- **Continuous integration and continuous delivery (CI/CD)**: Vertex AI can help with automating the process of building, testing, and deploying models, which is important for keeping models updated and making sure they are always running smoothly in production

- **Scalability**: Vertex AI can help with scaling ML models to handle large amounts of data and traffic, which is essential for maintaining the performance of models in production

- **Monitoring**: Vertex AI can help with monitoring and measuring the performance of ML models in production, which is vital for understanding how well models are working and identifying areas for improvement

Overall, Vertex AI can help with MLOps by providing a comprehensive platform for managing the entire ML life cycle on GCP, from development to deployment and maintenance, making it easier to build, deploy, and manage ML models in production.

This table shows which Vertex AI tools or features can help you implement which components of a typical MLOps pipeline:

| MLOps Components | Vertex AI Tool | Other GCP Tools |
| --- | --- | --- |
| Feature Management | Vertex AI Feature Store | N/A |
| Data Management | Vertex AI Datasets | BigQuery<br><br>Google Cloud Storage |
| Data Exploration & Analysis | Vertex AI Workbench | Data Fusion |
| Metadata Store | Vertex AI Metadata Store | N/A |
| Workflow Orchestration | Vertex AI Pipelines | Kubeflow on Google Kubernetes Engine (GKE)<br><br>Composer (Airflow) |
| Model Registry | Vertex AI Model Registry | N/A |
| Model Development | Vertex AI Training<br><br>Vertex AI Experiments | N/A |
| Model Serving/Prediction Service | Vertex AI Batch Predictions<br><br>Vertex AI Endpoints | Custom deployments on GCE or GKE |
| Monitoring | Vertex AI Monitoring | N/A |

Table 2.1 – MLOps to GCP product mapping

Let's look at these Vertex AI tools in detail.

**Vertex AI Workbench** is a fully managed Jupyter Notebook-based development environment that supports the entire data science workflow. The key features include the following:

- Integration with most Google Cloud data sources, such as BigQuery and Google Cloud Storage
- The ability to create and use highly scalable Jupyter **virtual machine** (**VM**) instances
- The ability to trigger and utilize most GCP services through a **software development kit** (**SDK**)
- The ability to kick off cluster and job creation in Dataproc (managed Spark service)
- Notebook scheduling

The following screenshot shows the Vertex AI Workbench dashboard with a list of all deployed Jupyter Notebook instances:

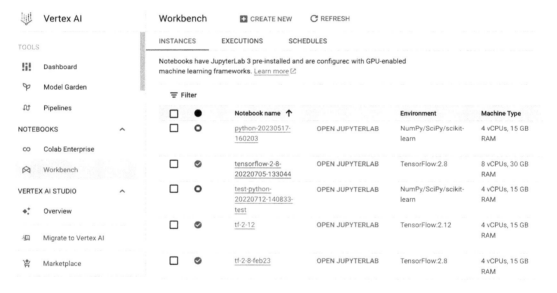

Figure 2.5 – Vertex AI Workbench dashboard

**Vertex AI Data Management** is, as the name suggests, Vertex AI's native dataset management service. It allows users to do the following:

- Import data into Vertex AI and manage the datasets
- Manage labels and multiple annotation sets

- When used in conjunction with Metadata Store, it helps track data lineage to models for troubleshooting and audits

- Generate data statistics and visualizations

The following screenshot shows the Vertex AI Datasets dashboard with a list of all datasets created with the tool:

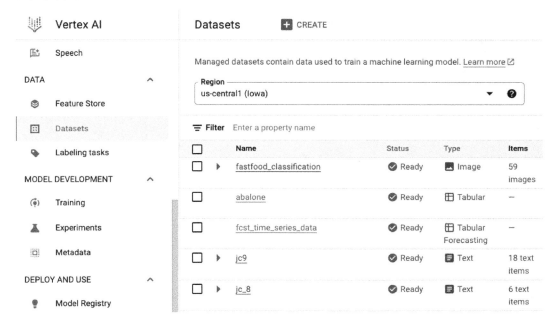

Figure 2.6 – Vertex AI Datasets dashboard

**Vertex AI Feature Store** is a managed feature catalog service, part of Vertex AI. Key features include the following:

- The ability to import and organize the feature values

- The ability to serve the feature values in batch or real-time modes

- The ability to monitor the change over time in feature values and generate alerts

The following screenshot shows the Vertex AI Feature Store dashboard with a list of all the features and entities present in the tool.

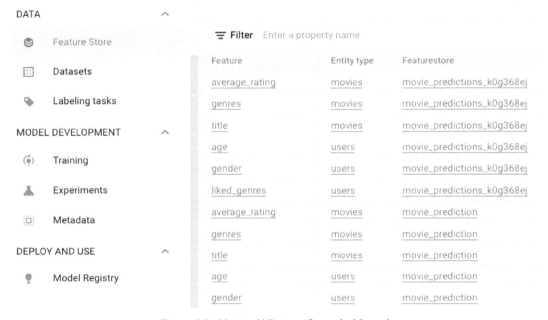

Figure 2.7 – Vertex AI Feature Store dashboard

**Vertex AI Pipelines** is a managed Kubeflow service in the Vertex AI platform, which helps you orchestrate complex data and ML workflows. Key features include the following:

- Support for the Kubeflow Pipelines SDK
- Support for **TensorFlow Extended** (**TFX**)
- The ability to deploy vertically scalable components as containers
- Native integration with all Vertex AI tools and most Google Cloud services

The following screenshot shows the Vertex AI Pipelines dashboard with a list of all recently executed pipelines.

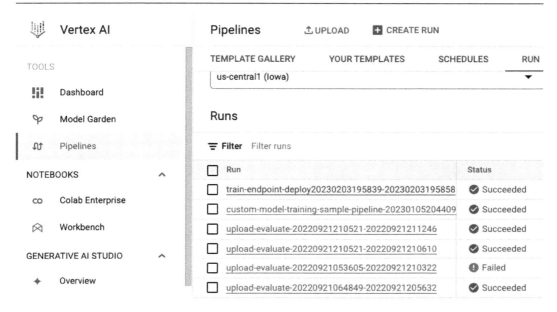

Figure 2.8 – Vertex AI Pipelines dashboard

**Vertex ML Metadata Store** can store key parameters/artifacts from the ML pipelines to enable the following:

- Lineage tracking and auditing for ML models

- Model reproducibility

- Analysis of experiments

- Tracking downstream usage of ML artifacts for better governance and compliance

The following screenshot shows a Vertex AI Metadata Store dashboard. The following figure shows a sample Metadata Store artifact depicting the lineage of an ML pipeline.

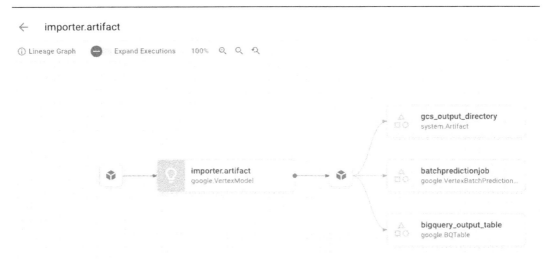

Figure 2.9 – Vertex AI Metadata Store

**Vertex AI Model Registry** is a centralized registry for all ML models regardless of whether they were custom models or models generated by using AutoML capabilities. Key features include the following:

- Fully managed model registry
- Seamless model deployments for a batch or real-time prediction service
- The ability to compare models

The following figure shows the Vertex AI Model Registry dashboard with a list of registered models.

| Model Registry | | | | | |
|---|---|---|---|---|---|
| tabnet_petfinder_classification_230831214806 | — | ⊙ Imported | Custom training | Aug 31, 2023, 2:48:10 PM | — |
| recommender_model_chp15 | ✓ Deployed | ⊙ Imported | Custom training | Aug 31, 2023, 3:24:37 AM | — |
| Fast_food_Classification_no_overlay | — | 🖼 Image classification | AutoML training | May 7, 2023, 2:25:21 AM | — |
| Bank_Marketing_Dataset | ✓ Deployed | ⊞ Tabular | AutoML training | May 6, 2023, 4:29:14 AM | — |
| Fast_food_Classification | ✓ Deployed | 🖼 Image classification | AutoML training | May 6, 2023, 3:32:51 AM | — |
| bank-20220621125144 | — | ⊞ Tabular | AutoML training | May 3, 2023, 3:32:50 PM | — |
| train_deploy20230203195839 | — | ⊙ Imported | Custom training | Feb 3, 2023, 12:40:21 PM | — |

Figure 2.10 – Vertex AI Model Registry

**Vertex AI Training** is the core managed training service in Vertex AI that enables users to run complex ML model training jobs without managing the underlying infrastructure. Key capabilities include the following:

- Native support for TensorFlow, XGBoost, and scikit-learn custom models
- The ability to deploy training jobs using custom containers
- Automate ML for a variety of use cases
- The ability to deploy models on extremely scalable on-demand clusters
- Managed TensorBoard

**Vertex AI Experiments** is a managed experiment orchestration feature that enables the data science team to kick off a number of similar training jobs with slightly differing hyperparameters as part of model design.

The following figure shows the Vertex AI Experiments dashboard with a list of executed training experiments.

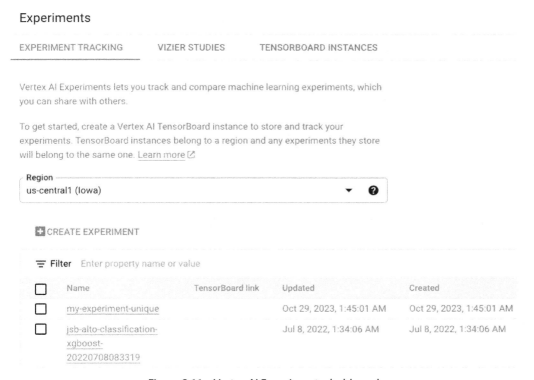

Figure 2.11 – Vertex AI Experiments dashboard

**Vertex AI Batch Predictions** provides fully managed batch predictions for models, uploaded to the model registry.

The following figure shows the Vertex AI Batch Predictions dashboard with a list of batch prediction jobs run on the platform.

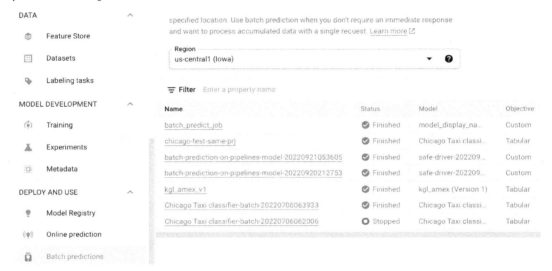

Figure 2.12 – Vertex AI Batch Predictions dashboard

**Vertex AI Endpoints** is a managed model serving capability for real-time prediction use cases. Key features include the following:

- Configurable serving infrastructure
- Autoscaling capabilities
- The ability to detect the number of performance issues relating to increased prediction latency, capacity bottlenecks, and so on

The following figure shows the dashboard for Vertex AI Endpoints.

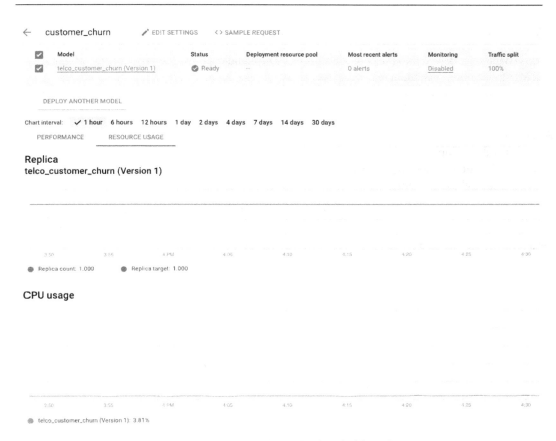

Figure 2.13 – Vertex AI monitoring dashboard

**Vertex AI Monitoring** helps automate the monitoring of deployed models in production to identify performance issues proactively. Key features include the following:

- Drift detection
- Training-serving skew detection

As you can see, Google Cloud Vertex AI offers a comprehensive and robust set of tools that can help you streamline the entire ML development life cycle, from data management to model deployment and monitoring.

## Summary

In this chapter, we delved into the core concepts of MLOps and explored the various levels of maturity that MLOps solutions exhibit in real-world scenarios. We also discussed the various tools and resources that are available as part of the Vertex AI platform, which are designed to help streamline and automate the process of building, deploying, and maintaining ML models.

After reading this chapter, we hope you are able to better articulate the fundamentals of MLOps practices and develop a high-level MLOps strategy for your organization, taking into consideration your organization's unique requirements and goals, and understand how Vertex AI, as a part of GCP, can play a crucial role in your organization's MLOps strategy.

In the next chapter, we cover the options available in Google Cloud to store data in order to support your ML solution's training needs. We also discuss the tools available to you to help you transform large-scale datasets as part of feature engineering activities.

# Part 2: Machine Learning Tools for Custom Models on Google Cloud

In this part, you will get an overview of the key ML tools that Google Cloud offers. You will learn about the different options in Google Vertex AI for storing and transforming data, training ML models, model optimization, model explainability, and so on. In addition to this, you will learn about the deployment tools, automation tools, and governance tools that Vertex AI offers for ML orchestration.

This part has the following chapters:

- *Chapter 3, It's All about Data – Options to Store and Transform ML Datasets*

- *Chapter 4, Vertex AI Workbench – a One-Stop Tool for AI/ML Development Needs*

- *Chapter 5, No-Code Options for Building ML Models*

- *Chapter 6, Low-Code Options for Building ML Models*

- *Chapter 7, Training Fully Custom ML Models with Vertex AI*

- *Chapter 8, ML Model Explainability*

- *Chapter 9, Model Optimizations – Hyperparameter Tuning and NAS*

- *Chapter 10, Vertex AI Deployment and Automation Tools – Orchestration through Managed Kubeflow Pipelines*

- *Chapter 11, MLOps Governance with Vertex AI*

# 3

# It's All About Data – Options to Store and Transform ML Datasets

The real work on a machine learning project only starts once the required data is available in the project development environment. Sometimes, when the data changes very frequently or the use case requires real-time data, we may need to set up some data pipelines to ensure that the required data is always available for analysis and modeling purposes. The best way to transfer, store, or transform data also depends on the size, type, and nature of the underlying data. Raw data, as collected in the real world, is often massive in size and may belong to multiple types, such as text, audio, images, videos, and so on. Due to the varying nature, size, and type of real-world data, it becomes really important to set up the correct infrastructure for storing, transferring, transforming, and analyzing the data at scale.

In this chapter, we will learn about the different options for moving data to the Google Cloud environment, different data storage systems, and efficient ways to apply transformations to large-scale data.

In this chapter, we will look at the following topics about data:

- Moving data to Google Cloud
- Where to store data
- Transforming data

# Moving data to Google Cloud

When we start a machine learning project on **Google Cloud Platform** (**GCP**), the very first step is to move all our project-related data to the Google Cloud environment. While transferring data to the cloud, the key things to focus on are reliability, security, scalability, and the ease of managing the transfer process. With these points in mind, Google Cloud provides four major data transfer utilities to meet customer requirements across a variety of use cases. In general, these utilities are useful for any kind of data transfer purposes, including data center migration, data backup, content storage, and machine learning. As our current focus is on making data available for machine learning use cases, we can utilize any of the following transfer solutions:

- Google Cloud Storage Transfer tools
- BigQuery Data Transfer Service
- Storage Transfer Service
- Transfer Appliance

Let's understand each of these transfer solutions.

## Google Cloud Storage Transfer tools

This option is suitable when our dataset is not too big (up to a few TB is fine), and we wish to store it in **Google Cloud Storage** (**GCS**) buckets (GCS is an object-type storage system, very similar to the local filesystem in our computers; we will learn more about it in the next section). Google Cloud provides tools for uploading data into these GCS buckets directly from our computers. We can upload one or more files or even a folder containing files using one of the following methods and also track the progress of uploads using the **upload progress** window from the Google Cloud console.

Here are the three methods for uploading files or folders to a GCS bucket:

- Using the Google Cloud console UI
- Using the command line
- Using the REST API (JSON API)

Let's look at these methods in more detail.

## Using the Google Cloud Console UI

It is quite easy to upload files or a folder containing files to GCS using the Cloud Console UI. When we upload a folder, the hierarchical structure inside the folder is also preserved. Follow these simple steps to upload data into a GCS bucket using the UI:

1.  Open a browser and go to the **Google Cloud Console** page.

2.  From the left pane, click on **Cloud Storage** and open the **Buckets** page. It will list all the existing buckets in our project.

3.  Click on the name of the relevant bucket if it already exists; otherwise, create a new bucket to store uploaded files or folders.

4.  Once we are inside the bucket, we will see bucket details and existing content. Now, we can directly upload the data using the **Upload Files** or **Upload Folder** button. The UI also provides options for creating a new folder and downloading or deleting files.

> **Note**
> Folder uploads are only supported using the Chrome browser. It may not work with other browsers.

## Using the command line

Google Cloud also provides an open source command-line utility called **GSUTIL**. We can utilize GSUTIL for scripted data transfers and also to manage our GCS buckets using simple commands. For large-scale streaming data, it supports multi-threaded/multi-processing data transfer for pushing script output. It can operate in *rsync* mode and transfer incremental copies of data.

GSUTIL commands are quite similar to Unix commands. See the following example for copying a file into the GCS bucket we created previously:

```
$ gsutil cp Desktop/file.txt gs://my-bucket-name
```

Similarly, we can list the content of a bucket using the `ls` command:

```
$ gsutil ls gs://my-bucket-name
$ gsutil ls gs://my-bucket-name/my-data-folder
```

### REST API (JSON API)

The JSON API interface lets us access or manipulate GCS data programmatically. This method is more suitable for software developers who are familiar with web programming and creating applications that consume web services using HTTP requests. For example, we can use the following HTTP request to list the objects of a particular bucket:

```
GET https://storage.googleapis.com/storage/v1/b/my-bucket/o
```

> **Important note**
>
> To use the preceding methods to access, manipulate, or upload data, we must have the proper **Identity and Access Management (IAM)** permissions. The project owner can provide a list of relevant permissions to the project development team.

## BigQuery Data Transfer Service

BigQuery Data Transfer Service currently supports loading data from Google **Software-as-a-Service (SaaS)** apps, external cloud providers, data warehouses such as **Teradata** or **Redshift**, and a few third-party sources. Once data is available, we can directly perform analytics or machine learning right inside **BigQuery (BQ)**. It can also be used as a data warehousing solution; we will learn more about BQ in the coming sections.

## Storage Transfer Service

Compared to GSUTIL, Storage Transfer Service is a managed service that is suitable for transferring data quickly and securely between object and file storage systems across different clouds (AWS and Azure), on-premises, or within different buckets in Google Cloud. The data transfer process is really fast as it utilizes a high bandwidth network. It also handles retries and provides detailed transfer logging.

## Transfer Appliance

This option is suitable when we want to migrate a really large dataset and don't have much bandwidth. Transfer Appliance is a physical device with a high memory capacity that can be utilized for transferring and securely shipping data to a Google upload facility, where data is uploaded to cloud storage. We can order the appliance from the Cloud Console, and once we receive the device, we can start copying our data. Finally, we can ship the appliance back to Google to transfer data into a specified GCS bucket.

For most machine learning-related use cases, the first two methods should be enough to transfer data fast, securely, and consistently. Next, let's learn more about the GCS and BQ data storage systems on Google Cloud.

# Where to store data

GCS and BQ are two recommended options for storing any machine learning use case-related datasets for high security and efficiency purposes. If the underlying data is structured or semi-structured, BQ is the recommended option due to its off-the-shelf features for manipulating or processing structured datasets. If the data contains images, videos, audio, and unstructured data, then GCS is the suitable option to store it. Let's learn about these two data storage systems in more detail.

## GCS – object storage

A significant amount of data that we collect from real-world applications is in unstructured form. Some examples are images, videos, emails, audio files, web pages, and sensor data. Managing and storing such huge amounts of unstructured data affordably and efficiently is quite challenging. Nowadays, object storage has become a preferable solution for storing such large amounts of static data and backups. Object storage is a computer data architecture that's designed to handle large amounts of structured data efficiently. Each data object in an object-based storage system is considered a distinct unit bundled with metadata and a unique identifier that is useful in quickly retrieving and locating data.

GCS is an object-based storage system in Google Cloud. As it is cloud-based, GCS data can be accessed globally and provides massive scale. It is a suitable option for small to large enterprises to store their large amounts of data in a cost-effective and easily retrievable fashion. Object storage is more efficient for applications where you write the data once but have to read it very frequently. While it is extremely good for static data, it's not a good solution for dynamic data. If data is constantly changing, we will have to write the entire data object again and again to modify it, which is inefficient.

GCS is frequently used by machine learning practitioners on Google Cloud due to its variety of benefits. Here are some common benefits of storing data in an object storage system such as GCS:

- **Massively scalable**: Object storage can be expanded infinitely by simply adding more servers or devices.
- **Less complex**: Unlike a file storage system, there is no hierarchy or folder structure in object storage, so retrieval is quite simple.
- **Searchability**: It is easy to search for a specific object as metadata is also part of the object. We can use tags to make objects more filterable.
- **Resiliency**: There's no fear of data loss as it can automatically replicate data and store it across multiple devices or geographical locations.
- **Low cost**: Object storage is cost-effective and thus ideal for storing large amounts of data. Secondly, we only pay for the capacity we use.

With so many advantages, there are also some limitations of object-based storage systems:

- Due to latency concerns, they cannot be used in place of traditional databases when designing web applications

- They're not suitable for situations when data is changing rapidly and lots of file writes are required very frequently

- They're not very compatible with operating system mounting and require additional clients or adapters to work with

Now that we have a fair idea about the advantages and limitations of object storage systems, we will be able to utilize them based on relevant requirements in future projects. Now, let's learn more about BQ.

## BQ – data warehouse

BQ is a fully managed and serverless data warehouse available on Google Cloud. It is a petabyte-scale platform that enables scalable analysis on large datasets. BQ's serverless architecture supports SQL queries for slicing and dicing large datasets. Its analysis engine is very scalable and supports distributed analysis such that we can query terabytes of data in seconds and petabytes in minutes. BQ also supports machine learning, meaning we can train and test common ML models within BQ using just a few SQL-like commands. We will learn about **BigQuery Machine Learning** (**BQML**) in the coming chapters.

Behind the scenes, BQ stores data in a columnar storage format that is optimized for analytical queries. Data inside BQ is presented in a database via tables with rows and columns. It provides full support for **Atomicity, Consistency, Isolation, and Durability** (**ACID**) properties, similar to a transactional database management system. BQ provides high availability for data by automatically replicating it across multiple locations and regions in Google Cloud. In addition to the data that is present inside BQ storage, it also provides flexibility to query data from external sources, including GCS, Bigtable, Google Sheets, and Spanner.

If a machine learning use case involves structured or semi-structured data, BQ can be the best place to store and analyze it. In subsequent chapters, we will learn more about how BQ is an extremely useful tool for data analysts and machine learning practitioners. Here are some common benefits of using BQ as a data warehousing solution:

- **Speed and scale**: With BQ, querying through massive datasets only takes seconds. BQ was designed to analyze and store very large amounts of datasets with ease. It can scale seamlessly from petabytes to exabytes.

- **Real-time analytics**: BQ supports streaming data ingestion and makes it immediately available for querying. Its integration with the BI tool Looker Studio allows it to provide real-time and interactive analytical capabilities.

- **Machine learning**: BQ provides capabilities to build and operationalize machine learning models on both structured and unstructured data using simple SQL – in a very short time. BQ models can be directly exported so that they can be served on the Vertex AI prediction service.

- **Security**: BQ provides strong security and fine-grained governance controls. Data is encrypted at rest and in transit by default.

- **Multi-cloud support**: BQ allows for data analysis across multiple cloud platforms. BQ can run analysis on data where it is located without having to move it to a different location, which makes it more cost-effective and flexible.

We now have a fair idea of two popular data storage systems – GCS and BQ – and their benefits and limitations. Depending on the use case, we should now be capable of choosing the right place to store our machine learning datasets. Next, we'll dig into data transformation options on Google Cloud.

# Transforming data

Raw data present in real-world applications is often unstructured and noisy. Thus, it cannot be fed directly to machine learning algorithms. We often need to apply several transformations on raw data and convert it into a format that is well supported by machine learning algorithms. In this section, we will learn about multiple options for transforming data in a scalable and efficient way on Google Cloud.

Here are three common options for data transformation in the GCP environment:

- Ad hoc transformation within Jupyter Notebooks
- Cloud Data Fusion
- Dataflow pipelines for scalable data transformations

Let's learn about these three methods in more detail.

## Ad hoc transformations within Jupyter Notebook

Machine learning algorithms are mathematical and can only understand numeric data. For example, in computer vision problems, images are converted into numerical pixel values before they're fed into a model. Similarly, in the case of audio data, it is often converted into a time-frequency domain using different transformations, such as **Fast Fourier Transformation** (FFT). If data is in a tabular format and contains rows and columns, some of the columns might contain non-numeric or categorical types of data. These categorical columns are first converted into numeric form using a suitable transformation and then fed to a machine learning model or neural network. Some of these transformations can be directly applied in Jupyter notebooks using Python functionalities.

After reading data into a Jupyter Notebook, we can apply any desired transformations to make the dataset ready for modeling. Once the dataset is ready, we can save it somewhere (for example, BQ or GCS) with a version number so that it can be read directly into multiple different model training experiments. Machine learning practitioners often create and save multiple versions of training datasets by applying different transformations or feature engineering. It makes it easier to compare the performance of experiments over different versions of data. Let's learn about some common transformations that are very frequently applied in machine learning projects.

At a high level, data can be classified into the following two categories:

- **Numeric data**: As the name suggests, this data is numeric or quantifiable – for example, age, height, and weight

- **Categorical data**: This data is in string format or is qualitative data – for example, gender, language, and accent

Now, let's learn about some common transformations that can be applied to these two types of data columns.

### Handling numeric data

Numerical data can either be discrete or continuous. Discrete data is countable, for example, the number of players in a basketball team or the number of cities in a country. Discrete data can only take certain values, such as 10, 22, 35, 41, and so on. On the other hand, any data that is measurable is called continuous data – for example, the height of a person or the distance between two racing cars. Continuous data can virtually take any value, such as 2.3, 11, 0.0001, and so on.

Two common kinds of transformations are applied to numeric data:

- **Normalizing**: This is done to bring different numeric columns to the same scale. It is recommended to normalize numerical data columns to the same scale as it helps in the convergence of gradient descent-based ML algorithms. If a data column has very large values, then normalizing it can prevent **NaN** errors while training models.

- **Bucketing**: In this type of transformation, numeric data is converted into categorical data. Bucketing is usually done on continuous numeric data when there is no linear relationship between the numeric column and the target column. For example, a car manufacturing company observes that the cars that have the lowest price or the highest price are less frequently sold, but the cars with mid-range prices are more frequently sold. In this case, if we want to predict the number of cars that are sold using car price as a feature, it will be beneficial to bucketize the car price into price ranges. In this way, the model will be able to identify the mid-range bucket in which most of the cars are sold and assign more weight to this feature.

## Handling categorical data

Categorical data can be divided into two categories – **ordered** and **nominal**. Ordered categorical data has some order associated with it – for example, movie ratings (worst, bad, good, excellent) and feedback (negative, neutral, positive). Ordered data can always be marked on a scale. Nominal data, on the other hand, has no order. Some examples of nominal data include gender, country, and language.

Some algorithms, such as decision trees, can work well with categorical data, but most machine learning algorithms cannot handle categorical data directly. These algorithms require the categorical data to be converted into numerical form. While converting the categorical data into numerical form, some challenges are faced by machine learning practitioners:

- **High cardinality**: Cardinality means uniqueness in data. A high cardinality data column might have lots of different values – for example, ZIP codes in country-level data.

- **Rare or frequent occurrences**: A data column might have some values that rarely occur or some values that occur very frequently. In both cases, this column would not be significant enough to make an impact on the model due to very high or very low variance.

- **Dynamic values**: A data column that keeps changing some values from time to time – for example, in a city column, if new cities are added or removed very frequently.

The best way to overcome these challenges highly depends on the kind of problem or data we are dealing with. Now, let's learn about some common methods of converting categorical data into numerical form:

- **Label encoding**: In this technique, we replace categorical data with integer values from 0 to $N$-1. Here, each integer represents a value from a categorical data column with $N$ unique values. For example, if there is a categorical data column that represents colors and 10 unique color values are possible, in this case, each color will be mapped and replaced with an integer from the range of 0 to 9. This method may not be ideal for all cases as the model might consider numeric values as weights assigned to the data. Thus, this method is more suitable for ordinal categorical data.

- **One-hot encoding**: In label encoding, data values are represented with integers (such as 0, 1, 2, and 3). Machine learning models might mistake these integer values to consider some kind of order. For example, a value encoded with the number 2 might be given two times the priority than another value that is encoded with the number 1, but this is a wrong assumption if values in data are not ordered (such as color names and city names). This issue with label encoding can be avoided using the one-hot encoding technique. In this method, each unique value from data is considered as a new binary column (here, *binary column* represents the presence or absence of that value in a given row). For example, if a color data column can have 10 different colors, it will be encoded into 10 new columns, each representing one color. The values in these columns will be 1 if that color is present in a given row; otherwise, it will be 0. One-hot encoding is often preferred for encoding categorical data.

- **Embeddings**: As one-hot encoding creates a new column for each unique value, this may create a very sparse representation of data when the number of unique values is large in number. For example, if we have a ZIP code column with 20k unique ZIP codes, the one-hot encoding method will create 20k new binary columns. Such sparse data takes a lot of memory to store and increases the complexity of machine learning training. To handle and represent such categorical data columns with a large number of unique values, dense embeddings can be used. These embeddings, however, are often generated using a neural network, so it's an off-the-shelf encoding technique. These embeddings encode each value from a categorical column into a small dense vector of real numbers. One simple method to train and generate these embeddings is using the inbuilt Keras embedding layer.

Now that we have a good understanding of common data transformations that are easy to apply in Jupyter notebooks using Python, let's learn about some more scalable ways of data transformation on GCP, such as Cloud Data Fusion and Dataflow.

## Cloud Data Fusion

Cloud Data Fusion is a fully managed service on GCP for quickly building and managing scalable data pipelines. Using the Data Fusion UI, we can build and deploy data pipelines without writing a single line of code (using a visual point-and-click interface). The Data Fusion UI lets us build scalable data integration solutions to clean, prepare, blend, transform, and transfer data in a fully managed way (which means we do not need to manage infrastructure). Cloud Data Fusion offers hundreds of prebuilt transformations for both batch and real-time data processing and quickly building ETL/ELT pipelines.

Some key features of Cloud Data Fusion are as follows:

- **Portability**: Cloud Data Fusion is built using an open source project called **CDAP**, thus ensuring data pipeline portability

- **Simple integration**: Cloud Data Fusion's easy integration with Google Cloud functionalities such as GCS, Dataproc, and BigQuery makes development faster and easier, ensuring security

- **No-code pipelines**: Even non-technical users can build data pipelines quickly using Cloud Data Fusion's web interface

- **Hybrid**: Since Cloud Data Fusion is an open source project, it provides flexibility to build standardized data pipelines across hybrid and multi-cloud environments

- **Security**: Cloud Data Fusion provides enterprise-grade security and access management with Google Cloud for data protection and compliance.

In addition to the web UI, we can also use command-line tools to create and manage Cloud Data Fusion instances and pipelines. Next, let's learn about another data transformation tool on GCP – Dataflow

## Dataflow pipelines for scalable data transformations

Dataflow is a managed service for executing data processing or transformation pipelines on Google Cloud that are developed using **Apache Beam SDK**. It supports unified batch and stream data processing that is fast, cost-effective, and serverless (which means we do not need to manage infrastructure). Because Dataflow is serverless, it lets us focus on expressing the business logic of our data pipeline (using SQL or code) without worrying about operational tasks and infrastructure management. Due to its streaming nature, Dataflow is ideal for building real-time pipelines for use cases such as anomaly detection, pattern recognition, and forecasting.

By combining Dataflow with other managed Google Cloud services, we can simplify many aspects of productionizing data pipelines compared to self-managed solutions. For example, it can be combined with Google Cloud offerings such as Pub/Sub and BQ to develop a streaming solution that can ingest, process, and analyze fluctuating volumes of real-time data and generate invaluable real-time business insights. As it is managed, it provisions and scales the required resources automatically and thus reduces the time and complexity for data engineers or data analysts working on stream analytics solutions.

Some key features of Dataflow are as follows:

- **Smart autoscaling**: Dataflow supports both horizontal and vertical scaling of worker nodes. Scaling is performed automatically in such a way that the utilization of worker nodes and other pipeline scaling requirements are met in an efficient (cost-effective) or best-fit manner.

- **Real-time pipelines**: Dataflow's streaming nature is useful in building real-time stream analytics, machine learning forecasting, and anomaly detection pipelines. It is also useful for synchronizing or replicating data across multiple data sources (such as BQ, PostgreSQL, or Cloud Spanner) with minimal latency.

- **Dataflow SQL**: Dataflow streaming pipelines can be built using simple SQL commands right from BQ.

- **Flexible scheduling**: Batch processing jobs that need scheduling (such as overnight jobs) can easily be scheduled using Dataflow **Flexible Resource Scheduling** (**FlexRS**) in a cost-effective setting.

# Summary

Managing data effectively is really important for saving time, cost, and complexity for every organization. A machine learning practitioner should be aware of the best options for transferring, storing, and transforming data to build machine learning solutions more efficiently. In this chapter, we learned about multiple ways of bringing data into the Google Cloud environment. We discussed the best options for storing it based on the characteristics of the data. Finally, we discussed multiple different tools and methods for transforming/processing data in a scalable manner.

After reading this chapter, you should feel confident about choosing the best option for moving or transferring data into your Google Cloud environment based on the requirements of the use case. Choosing the best place to store data and the best strategy to analyze and transform data should be easier as we now know the pros and cons of different options. In the next chapter, we will deep dive into **Vertex AI Workbench**, which is a managed notebook platform within Vertex AI.

# 4

# Vertex AI Workbench – a One-Stop Tool for AI/ML Development Needs

**Machine learning** (**ML**) projects are complex in nature and require an entirely different type of development environment from normal software applications. When the data is huge, a data scientist may want to use several big data tools for quick wrangling or preprocessing needs, and a **deep learning** (**DL**) model might require several GPUs for fast training and experimentation. Additionally, dedicated compute resources are required for hosting models in production, and even more to scale them up to the enterprise level. Acquiring such resources and tools is quite costly, and even if we manage to buy and set things up, it takes a lot of effort and technical knowledge to bring them together into a project pipeline. Even after doing all that, there are risks of downtime and data security.

Nowadays, cloud-based solutions are very popular and take care of all the technical hassle, scaling, and security aspects for us. These solutions let ML developers focus more on project development and experimentation without worrying about infrastructure and other low-level things. As an **artificial intelligence** (**AI**)-first company, Google brings all the important resources required for ML project development under one umbrella called Vertex AI. In this chapter, we will learn about Vertex AI Workbench, a managed solution for Jupyter Notebook kernels that can help us bring our ML projects from prototype to production many times faster.

This chapter covers the following topics:

- What is Jupyter Notebooks?
- Vertex AI Workbench
- Custom containers for Vertex AI Workbench
- Scheduling notebooks in Vertex AI

# What is Jupyter Notebook?

Jupyter Notebook is an open source web-based application for writing and sharing live code, documentation, visualizations, and so on. Jupyter Notebooks are very popular among ML practitioners as they provide the flexibility to run code dynamically and collaborate, provide fast visualizations, and can also be used for presentations. Most data scientists and ML practitioners prefer Jupyter Notebook as their primary tool for exploring, visualizing, and preprocessing data using powerful Python libraries such as pandas and NumPy. Jupyter Notebooks are very useful for **exploratory data analysis** (**EDA**) as they let us run small code blocks dynamically and also draw quick plots to understand data statistically. Notebooks can also be used for doing quick ML modeling experiments. Another good thing about Jupyter Notebooks is that they let us write Markdown cells as well. Using Markdown, we can explain each code block inside the notebook and turn it into a tutorial. Jupyter Notebooks are popular among ML communities to share and collaborate on projects on platforms such as GitHub and Kaggle.

## Getting started with Jupyter Notebook

The Jupyter Notebook application can be installed in local systems using a simple `pip` command (shown next). For quick experiments, we can also utilize web-based notebook kernels such as Colab and Kaggle, where everything is already set and we can run the Python code directly. As these kernels are public, we can't use them if our data is confidential, and we will have to install the Jupyter Notebook application on our system.

We can install the Jupyter application on our local system by using the following `pip` command:

```
$ pip install jupyter
```

Once the application is installed, it can be launched through the terminal by typing the following command, and it will automatically open the Jupyter application in a browser tab:

```
$ jupyter notebook
```

If it doesn't open the browser tab automatically, we can launch the application by typing the following URL: `http://localhost:8888/tree`. By default, the Jupyter server starts on port `8888`, but if this port is unavailable, it finds the next available port. If we are interested in using a custom port, we can launch Jupyter by passing a custom port number.

Here is a terminal command for launching the Jupyter application on custom port number `9999`:

```
$ jupyter notebook --port 9999
```

> **Note**
>
> In some cases, the Jupyter server may ask for a token (maybe in the case of a non-default browser) when we try to hit the aforementioned URL manually. In such cases, we can copy the URL from the terminal output that provides the token within the URL. Alternatively, we can obtain a token by running the `jupyter notebook list` command in the terminal.

Once we are able to launch the application server in a browser, the Jupyter server looks something like this:

Figure 4.1 – Jupyter application server UI

Now, we can launch a Jupyter Notebook instance by clicking on the **New** button. It creates a new notebook and saves it in the same directory where we started the Jupyter Notebook from the terminal. We can now open that notebook in a new tab and start running scripts. The following screenshot shows an empty notebook:

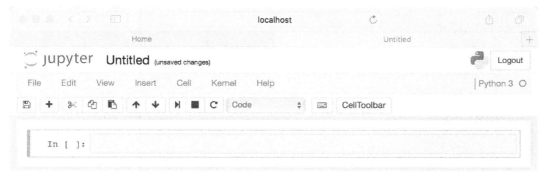

Figure 4.2 – A Jupyter Notebook instance

As we can see in the previous screenshot, the web UI provides multiple options to manipulate notebooks, code, cells, kernels, and so on. A notebook cell can execute code or can be converted into a Markdown cell by changing its type from the drop-down menu. There are also options for exporting notebooks into different formats such as HTML, PDF, Markdown, LaTeX, and so on for creating reports or presentations. Going further in the book, we will be working with notebooks a lot for data wrangling, modeling, and so on.

Now that we have some basic understanding of Jupyter Notebooks in general, let's see how Vertex AI Workbench provides a more enriched experience of working with a Jupyter Notebook-based environment.

## Vertex AI Workbench

While working on an ML project, if we are running a Jupyter Notebook in a local environment, or using a web-based Colab- or Kaggle-like kernel, we can perform some quick experiments and get some initial accuracy or results from ML algorithms very fast. But we hit a wall when it comes to performing large-scale experiments, launching long-running jobs, hosting a model, and also in the case of model monitoring. Additionally, if the data related to a project requires some more granular permissions on security and privacy (fine-grained control over who can view/access the data), it's not feasible in local or Colab-like environments. All these challenges can be solved just by moving to the cloud. Vertex AI Workbench within Google Cloud is a JupyterLab-based environment that can be leveraged for all kinds of development needs of a typical data science project. The JupyterLab environment is very similar to the Jupyter Notebook environment, and thus we will be using these terms interchangeably throughout the book.

Vertex AI Workbench has options for creating *managed notebook instances* as well as *user-managed notebook instances*. User-managed notebook instances give more control to the user, while managed notebooks come with some key extra features. We will discuss more about these later in this section. Some key features of the Vertex AI Workbench notebook suite include the following:

- Fully managed–Vertex AI Workbench provides a Jupyter Notebook-based fully managed environment that provides enterprise-level scale without managing infrastructure, security, and user-management capabilities.

- Interactive experience–Data exploration and model experiments are easier as managed notebooks can easily interact with other Google Cloud services such as storage systems, big data solutions, and so on.

- Prototype to production AI–Vertex AI notebooks can easily interact with other Vertex AI tools and Google Cloud services and thus provide an environment to run end-to-end ML projects from development to deployment with minimal transition.

- Multi-kernel support–Workbench provides multi-kernel support in a single managed notebook instance including kernels for tools such as TensorFlow, PyTorch, Spark, and R. Each of these kernels comes with pre-installed useful ML libraries and lets us install additional libraries as required.

- Scheduling notebooks–Vertex AI Workbench lets us schedule notebook runs on an ad hoc and recurring basis. This functionality is quite useful in setting up and running large-scale experiments quickly. This feature is available through managed notebook instances. More information will be provided on this in the coming sections.

With this background, we can now start working with Jupyter Notebooks on Vertex AI Workbench. The next section provides basic guidelines for getting started with notebooks on Vertex AI.

## Getting started with Vertex AI Workbench

Go to the Google Cloud console and open **Vertex AI** from the products menu on the left pane or by using the search bar on the top. Inside Vertex AI, click on **Workbench**, and it will open a page very similar to the one shown in *Figure 4.3*. More information on this is available in the official documentation (`https://cloud.google.com/vertex-ai/docs/workbench/introduction`).

Figure 4.3 – Vertex AI Workbench UI within the Google Cloud console

As we can see, Vertex AI Workbench is basically Jupyter Notebook as a service with the flexibility of working with managed as well as user-managed notebooks. User-managed notebooks are suitable for use cases where we need a more customized environment with relatively higher control. Another good thing about user-managed notebooks is that we can choose a suitable Docker container based on our development needs; these notebooks also let us change the type/size of the instance later on with a restart.

To choose the best Jupyter Notebook option for a particular project, it's important to know about the common differences between the two solutions. *Table 4.1* describes some common differences between fully managed and user-managed notebooks:

| Vertex AI-managed notebooks | Vertex AI user-managed notebooks |
| --- | --- |
| Google-managed environment with integrations and features that provide us with an end-to-end notebook-based production environment without setting up anything by hand. | Heavily customizable VM instances (with prebuilt DL images) that are ideal for users who need a lot of control over the environment. |
| Scaling up and down (for vCPUs and RAM) can be performed from within the notebook itself without needing to restart the environment. | Changing the size/memory of an instance requires stopping the instance in the Workbench UI and restarting it every time. |
| Managed notebooks let us browse data from **Google Cloud Storage** (**GCS**) and BigQuery without leaving the Jupyter environment (with GCS and BigQuery integrations). | UI-level data browsing is not supported in user-managed notebooks. However, we can read the data using Python in a notebook cell and view it. |
| Automated notebook runs are supported with one-time and recurring schedules. The executor runs scheduled tasks and saves results even when an instance is in a shutdown state. | Automated runs are not yet supported in a user-managed environment. |
| Less control over networking and security. | Option to implement desired networking and security features and VPC service controls on a per-need basis. |
| Not much control for a DL-based environment while setting up notebooks. | User-managed instances provide multiple DL VM options to choose from during notebook creation. |

Table 4.1 – Differences between managed and user-managed notebook instances

Let's create one user-managed notebook to check the available options:

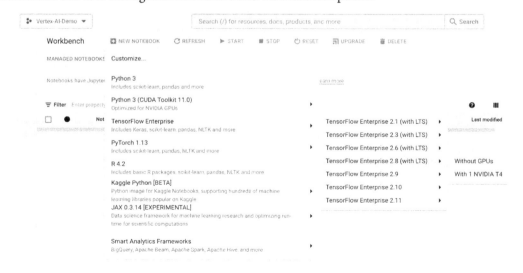

Figure 4.4 – Jupyter Notebook kernel configurations

As we can see in the preceding screenshot, user-managed notebook instances come with several customized image options to choose from. Along with the support of tools such as TensorFlow Enterprise, PyTorch, JAX, and so on, it also lets us decide whether we want to work with GPUs (which can be changed later, of course, as per needs). These customized images come with all useful libraries pre-installed for the desired framework, plus provide the flexibility to install any third-party packages within the instance.

After choosing the appropriate image, we get more options to customize things such as notebook name, notebook region, operating system, environment, machine types, accelerators, and so on (see the following screenshot):

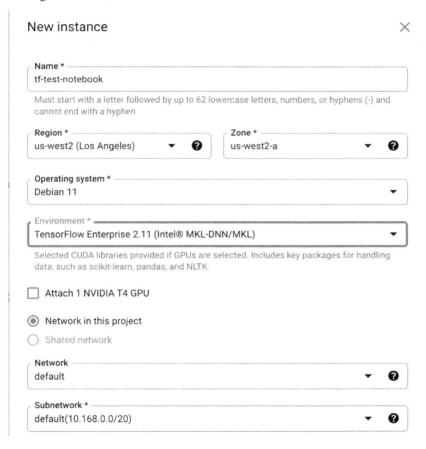

Figure 4.5 – Configuring a new user-managed Jupyter Notebook

Once we click on the **CREATE** button, it can take a couple of minutes to create a notebook instance. Once it is ready, we can launch the Jupyter instance in a browser tab using the link provided inside Workbench (see *Figure 4.6*). We also get the option to stop the notebook for some time when we are not using it (to reduce cost):

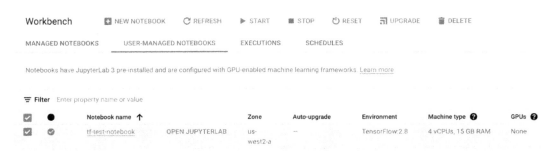

Figure 4.6 – A running Jupyter Notebook instance

This Jupyter instance can be accessed by all team members having access to Workbench, which helps in collaborating and sharing progress with other teammates. Once we click on **OPEN JUPYTERLAB**, it opens a familiar Jupyter environment in a new tab (see *Figure 4.7*):

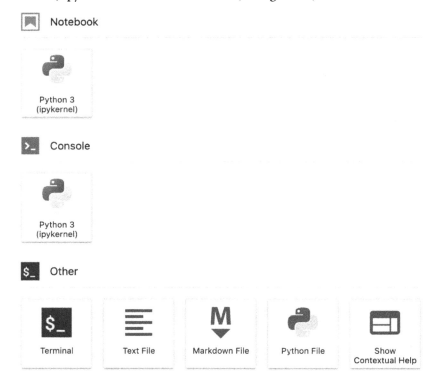

Figure 4.7 – A user-managed JupyterLab instance in Vertex AI Workbench

A Google-managed JupyterLab instance also looks very similar (see *Figure 4.8*):

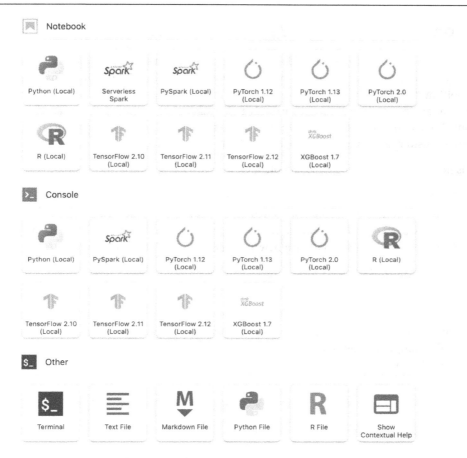

Figure 4.8 – A Google-managed JupyterLab instance in Vertex AI Workbench

Now that we can access the notebook instance in the browser, we can launch a new Jupyter Notebook or terminal and get started on the project. After providing sufficient permissions to the service account, many useful Google Cloud services such as BigQuery, GCS, Dataflow, and so on can be accessed from the Jupyter Notebook itself using SDKs. This makes Vertex AI Workbench a one-stop tool for every ML development need.

> **Note**
>
> We should stop Vertex AI Workbench instances when we are not using them or don't plan to use them for a long period of time. This will help prevent us from incurring costs from running them unnecessarily for a long period of time.

In the next sections, we will learn how to create notebooks using custom containers and how to schedule notebooks with Vertex AI Workbench.

# Custom containers for Vertex AI Workbench

Vertex AI Workbench gives us the flexibility of creating notebook instances based on a custom container as well. The main advantage of a custom container-based notebook is that it lets us customize the notebook environment based on our specific needs. Suppose we want to work with a new TensorFlow version (or any other library) that is currently not available as a predefined kernel. We can create a custom Docker container with the required version and launch a Workbench instance using this container. Custom containers are supported by both managed and user-managed notebooks.

Here is how to launch a user-managed notebook instance using a custom container:

1. The first step is to create a custom container based on the requirements. Most of the time, a derivative container (a container based on an existing DL container image) would be easy to set up. See the following example Dockerfile; here, we are first pulling an existing TensorFlow GPU image and then installing a new TensorFlow version from the source:

   ```
   FROM gcr.io/deeplearning-platform-release/tf-gpu:latest

   RUN pip install -y tensorflow
   ```

2. Next, build and push the container image to **Container Registry**, such that it should be accessible to the **Google Compute Engine** (**GCE**) service account. See the following source to build and push the container image:

   ```
   export PROJECT=$(gcloud config list project
   --format     "value(core.project)")

   docker build . -f Dockerfile.example -t "gcr.io/${PROJECT}/
   tf-custom:latest"

   docker push "gcr.io/${PROJECT}/tf-custom:latest"
   ```

   Note that the service account should be provided with sufficient permissions to build and push the image to the container registry, and the respective APIs should be enabled.

3. Go to the **User-managed notebooks** page, click on the **New Notebook** button, and then select **Customize**. Provide a notebook name and select an appropriate **Region** and **Zone** value.

4. In the **Environment** field, select **Custom Container**.

5. In the **Docker Container Image** field, enter the address of the custom image; in our case, it would look like this:

   ```
   gcr.io/${PROJECT}/tf-custom:latest
   ```

6. Make the remaining appropriate selections and click the **Create** button.

We are all set now. While launching the notebook, we can select the custom container as a kernel and start working on the custom environment.

We can now successfully launch Vertex AI notebooks and also create custom container-based environments if required. In the next section, we will learn how to schedule notebook runs within Vertex AI.

# Scheduling notebooks in Vertex AI

Jupyter Notebook environments are great for doing some initial experiments. But when it comes to launching long-running jobs, multiple training trials with different input parameters (such as hyperparameter tuning jobs), or adding accelerators to training jobs, we usually copy our code into a Python file and launch experiments using custom Docker containers or managed pipelines such as Vertex AI pipelines. Considering this situation and to minimize the duplication of efforts, Vertex AI-managed notebook instances provide us with the functionality of scheduling notebooks on an ad hoc or recurring basis. This feature allows us to execute our scheduled notebook cell by cell on Vertex AI. It provides us with the flexibility to seamlessly scale our processing power and choose suitable hardware for the task. Additionally, we can pass different input parameters for experimentation purposes.

## Configuring notebook executions

Let's try to configure notebook executions to check the various options it provides. Imagine we are building a toy application that takes two parameters–user_name and frequency–and when executed, it prints the user_name parameter as many times as the frequency parameter. Now, let's launch a managed notebook and create our application, as follows:

Figure 4.9 – A simple Python application within Jupyter Notebook

Next, put all the parameters into a single cell and click on the gear-like button at the top-right corner. Assign this cell with tag *parameters*. See the following screenshot:

Figure 4.10 – Tagging parameters within a Jupyter Notebook cell

Our toy application is now ready. Once you click on the **Execute** button from the toolbar, it provides us with the options for customizing machine type, accelerators, environment (which can be a custom Docker container), and execution type–one-time or recurring. See the following screenshot:

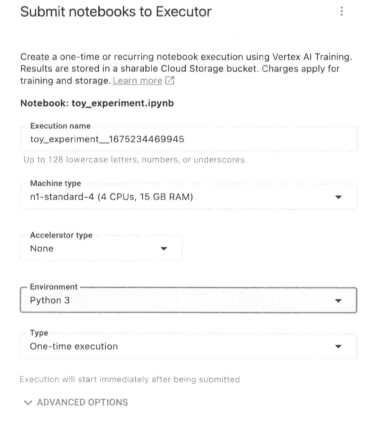

Figure 4.11 – Configuring notebook execution for Python application

Next, let's change the parameters for our one-time execution by clicking on the **ADVANCED OPTIONS** Here, we can provide key-value pairs for parameter names and values. Check the following screenshot:

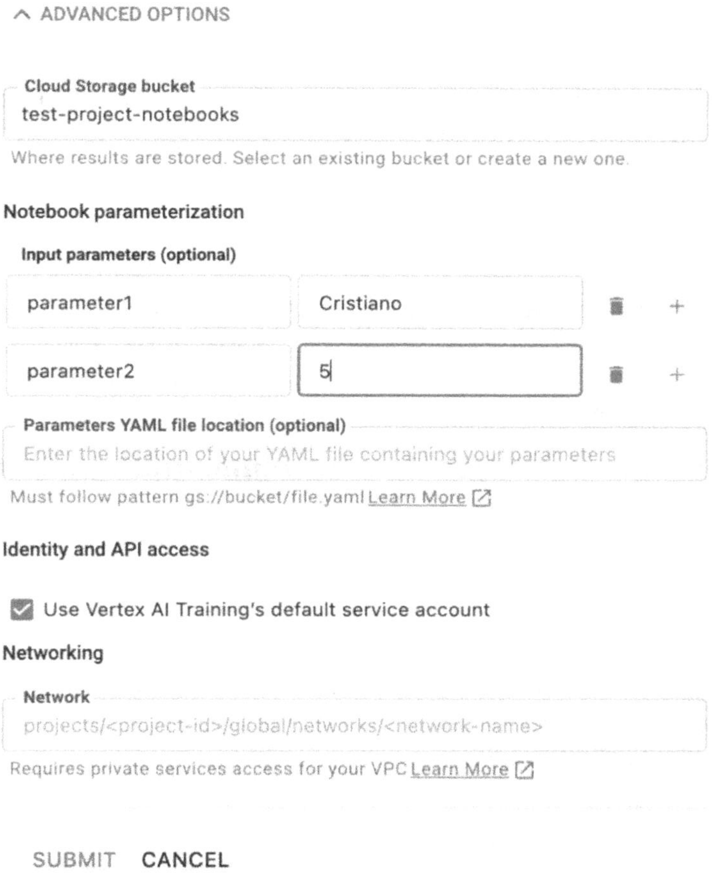

∧ ADVANCED OPTIONS

Cloud Storage bucket
test-project-notebooks

Where results are stored. Select an existing bucket or create a new one.

**Notebook parameterization**

**Input parameters (optional)**

| parameter1 | Cristiano | 🗑 | + |
| parameter2 | 5 | 🗑 | + |

Parameters YAML file location (optional)
Enter the location of your YAML file containing your parameters

Must follow pattern gs://bucket/file.yaml Learn More ☑

**Identity and API access**

☑ Use Vertex AI Training's default service account

**Networking**

Network
projects/<project-id>/global/networks/<network-name>

Requires private services access for your VPC Learn More ☑

SUBMIT    CANCEL

Figure 4.12 – Setting up parameters for one-time execution

Finally, click the **SUBMIT** button. It will then display the following dialog box:

 **One-time execution created!**

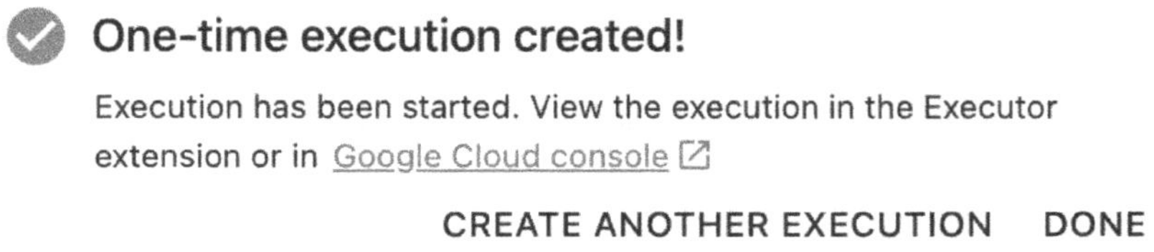

Execution has been started. View the execution in the Executor extension or in Google Cloud console ☑

CREATE ANOTHER EXECUTION    DONE

Figure 4.13 – One-time execution scheduled

We have now successfully scheduled our notebook run with custom parameters on Vertex AI. We can find it under the **EXECUTIONS** section in the Vertex AI UI:

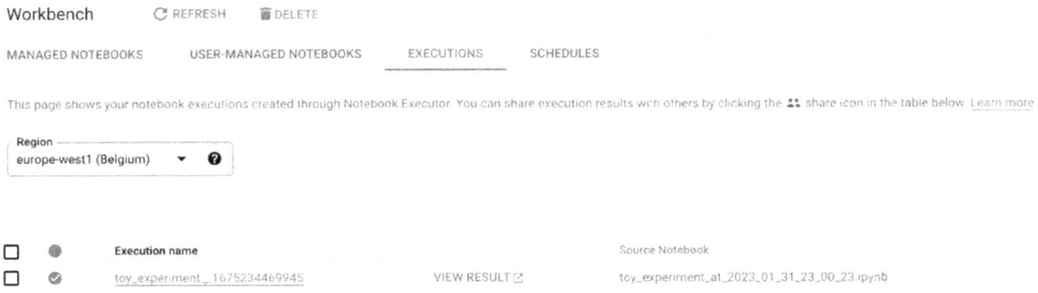


Figure 4.14 – Checking the EXECUTIONS section for executed notebook instances

We can now check the results by clicking on **VIEW RESULT**. Check the following screenshot for how it overrides the input parameters:

# This toy experiment prints user_name multiple times¶

In [1]:

```
user_name = 'Dummy User'
frequency = 2
```

In [2]:

```
# Parameters
user_name = "Cristiano"
frequency = "5"
```

In [3]:

```
for iteration in range(int(frequency)):
    print("User name is - {}, iteration - {}".format(user_name, iteration))
```

```
User name is - Cristiano, iteration - 0
User name is - Cristiano, iteration - 1
User name is - Cristiano, iteration - 2
User name is - Cristiano, iteration - 3
User name is - Cristiano, iteration - 4
```

In [ ]:

Figure 4.15 – Checking the results of the execution

Similarly, we can schedule large one-time or recurring experiments without moving our code out of the notebook and take advantage of the cloud platform's scalability.

We just saw how easy it is to configure and schedule notebook runs within Vertex AI Workbench. This capability allows us to do seamless experiments while keeping our code in the notebook. This is also helpful in setting up recurring jobs in the development environment.

## Summary

In this chapter, we learned about Vertex AI Workbench, a managed platform for launching the Jupyter Notebook application on Google Cloud. We talked about the benefits of having notebooks in a cloud-based environment as compared to a local environment. Having Jupyter Notebook in the cloud makes it perfect for collaboration, scaling, adding security, and launching long-running jobs. We also discussed additional features of Vertex AI Workbench that are pretty useful while working on different aspects of ML project development.

After reading this chapter, we should be able to successfully deploy, manage, and use Jupyter Notebooks on the Vertex AI platform for our ML development needs. As we understand the difference between managed and user-managed notebook instances, we should be in good shape to choose the best solution for our development needs. We should also be able to create custom Docker container-based notebooks if required. Most importantly, we should now be able to schedule notebook runs for recurring as well as one-time execution based on the requirements. Notebook scheduling is also quite useful for launching multiple model training experiments in parallel with different input parameters. Now that we have a good background in Vertex AI Workbench, it will be easier for us to follow the code samples in the upcoming chapters.

# 5

# No-Code Options
# for Building ML Models

In recent years, the world of machine learning has undergone a profound transformation, breaking free from the realm of expert data scientists and engineers to empower a broader audience. The rise of no-code machine learning platforms has ushered in a new era, where individuals with diverse skill sets and backgrounds can harness the power of artificial intelligence to solve complex challenges, without writing a single line of code. This democratization of machine learning has not only expedited the development process but has also opened up a myriad of opportunities for businesses and individuals alike.

In this chapter, we will dive into the foundations of no-code machine learning, shedding light on the remarkable tools and services Google Cloud Vertex AI offers. We will explore how users can leverage prebuilt machine learning models, AutoML capabilities, and visual interfaces to construct sophisticated and highly accurate models with ease. From computer vision to natural language processing and tabular data analysis, Google Cloud Vertex AI covers a vast array of use cases, democratizing AI application development for everyone.

The key topics we will cover in this chapter include the following:

- What is AutoML?
- What is Vertex AI AutoML?
- Creating and deploying a model using Vertex AI AutoML
- Getting predictions from a deployed Vertex AI model

Let's first start by looking at the different solutions offered by Google Cloud to facilitate model creation without using code.

# ML modeling options in Google Cloud

Google Cloud offers several solutions within Vertex AI and the broader **Google Cloud Platform** (**GCP**) to build and consume machine learning models. These solutions vary widely in terms of required data science and coding skills, catering to both advanced ML engineers, relatively less technical business analysts, and everyone in between these two personas. The three main GCP solutions for model creation are as follows:

- **Big Query ML (BQML)**: This is part of the BigQuery platform and requires only the knowledge of SQL for someone to train and use a model to generate predictions on structured data. More details about BQML will be covered in *Chapter 6, Low-Code Options for Building ML Models*.

- **Vertex AI AutoML**: This allows users to build models with no coding or even SQL knowledge, and it is primarily GUI-based. However, it has APIs that can be accessed programmatically if required.

- **Vertex AI custom training**: This option provides users complete flexibility on their model training and deployment but also requires basic coding ability.

The following table shows a comparison of the different options available in Google Cloud to create machine learning models:

| | BQML | Vertex AI AutoML | Vertex AI custom |
|---|---|---|---|
| Coding requirements | Very Low | None | High |
| Required ML engineering expertise | Low | Low | Medium to high, depending on the type of model |
| Limits on data size | Yes. Standard BigQuery quotas and limits apply. | Yes. Dataset limitations vary by the type of dataset being used (see the GCP documentation for current limits). | No limit is imposed by GCP for datasets that are not managed. |

| | BQML | Vertex AI AutoML | Vertex AI custom |
|---|---|---|---|
| Types of models supported | <ul><li>LINEAR REG</li><li>LOGISTIC REG</li><li>KMEANS</li><li>MATRIX FACTORIZATION</li><li>PCA</li><li>AUTOENCODER</li><li>AUTOML CLASSIFIER</li><li>AUTOML REGRESSOR</li><li>BOOSTED TREE CLASSIFIER</li><li>BOOSTED TREE REGRESSOR</li><li>RANDOM FOREST CLASSIFIER</li><li>RANDOM FOREST REGRESSOR</li><li>DNN CLASSIFIER</li><li>DNN REGRESSOR</li><li>DNN LINEAR COMBINED CLASSIFIER</li><li>DNN LINEAR COMBINED REGRESSOR</li><li>ARIMA PLUS</li><li>ARIMA PLUS XREG</li><li>TENSORFLOW</li><li>TENSORFLOW LITE</li><li>ONNX</li><li>XGBOOST</li></ul> | **Image**:<ul><li>Classification</li><li>Object detection</li></ul>**Text**:<ul><li>Classification</li><li>Entity extraction</li><li>Sentiment analysis</li></ul>**Video**:<ul><li>Action recognition</li><li>Classification</li><li>Object tracking</li></ul>**Tabular**:<ul><li>Regression</li><li>Classification</li><li>Time series forecasting</li></ul> | Full flexibility to build any type of ML model |

| | BQML | Vertex AI AutoML | Vertex AI custom |
|---|---|---|---|
| Model development speed | Fast. Some data preparation is required, but training can be mostly automated. | Fast. Minimal data preparation and fully automated model training. | Slower. More data preparation is required. Significant model design and training management. |
| Flexibility/control over model generation | Medium | Low | High |
| Does the tool support feature engineering? | Yes | No | Yes |

Table 5.1 – ML model creation options in Google Cloud

As shown in the preceding table, Vertex AI AutoML is the key code-less model creation option available as part of Google Cloud. When AutoML was initially launched as part of GCP, it used to be a standalone product. Now, it is part of the overall Vertex AI platform, and the legacy AutoML product is on the roadmap to be sunset. Now, let's understand what AutoML is and how you can use AutoML features available in GCP.

## What is AutoML?

AutoML refers to the methodology of automating the process of building machine learning models, including data preprocessing, feature engineering, model selection, hyperparameter tuning, and model deployment. AutoML aims to make machine learning accessible and more efficient for non-experts, saving time and resources for experts by reducing the amount of manual work involved in building a model. Different types of AutoML products on the market offer different levels of automation. Some just automate the training and hyperparameter portion of it, while some do end-to-end automation by also automating the steps of data preprocessing and feature generation.

AutoML tools allow users to specify their requirements, such as accuracy, interpretability, or training time, and then automatically select and train the best model based on these criteria. It can be used for various types of machine learning tasks, including classification, regression, and time series forecasting on structured and unstructured datasets. AutoML technologies have seen rapid development in recent years and are extremely capable of handling many complex ML use cases now, with minimal human intervention. However, you need to be careful about being overly dependent on AutoML. It is not a substitute for a deep understanding of machine learning and data science but, rather, a tool that can help make these processes more efficient and accessible. AutoML tools can be risky when used by someone who does not understand the fundamentals of machine learning because they can generate seemingly high-performing models, while still suffering from common issues such as data leakages and overfitting. So, now that we have a basic understanding of the concept of AutoML as it relates to model development, let's look at the AutoML features available in Vertex AI.

# Vertex AI AutoML

AutoML tools in Google Cloud have existed a lot longer than Vertex AI, which was launched to primarily unify most of the separate ML offerings existing in GCP. GCP AutoML makes use of models such as NASNet and constantly benefits from the AI research happening in other divisions of the Alphabet teams, such as Google DeepMind. Few of the interesting papers on the topic are listed below:

- *Learning Transferable Architectures for Scalable Image Recognition.* `https://arxiv.org/pdf/1707.07012.pdf`

- *Regularized Evolution for Image Classifier Architecture Search.* `https://arxiv.org/pdf/1802.01548.pdf`

- *Large-Scale Evolution of Image Classifiers.* `https://arxiv.org/pdf/1703.01041.pdf`

The use cases supported by Vertex AI AutoML are as follows:

- Tabular data:

  - Classification (an example is covered in this chapter)

  - Regression

  - Time series forecasting

- Image data:

  - Image classification (an example is covered in *Chapter 8*)

  - Object detection

- Natural language:

  - Text classification

  - Entity extraction

  - Sentiment analysis

## How to create a Vertex AI AutoML model using tabular data

In the following use case, we will walk you through the steps of building a classification model, using a public dataset containing hotel reservation data. The model's objective will be to predict the probability of a particular hotel reservation being canceled by the customer, helping the hotel to better plan around future room occupancy and possibly allow for overbooking in the hotel on dates where they expect a high number of cancellations:

* The hotel reservation dataset can be accessed here: `https://www.kaggle.com/datasets/ jessemostipak/hotel-booking-demand`

* You can download the data from the GitHub repository accompanying this book: `https:// github.com/PacktPublishing/The-Definitive-Guide-to-Google- Vertex-AI`

# Importing data to use with Vertex AI AutoML

The first step when planning to use the Vertex AI AutoML feature is to import the data you plan to use to train as Vertex AI datasets:

1. Navigate to **Vertex AI | Datasets** within the Google Cloud console, and click **Create** to start creating a new Vertex AI dataset.

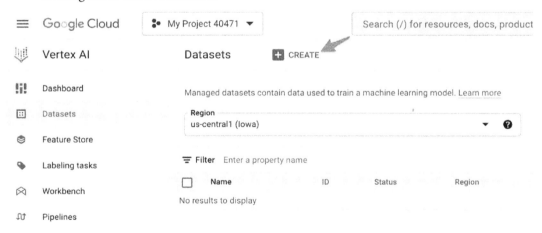

Figure 5.1 – Creating a Vertex AI dataset

2. Type in the name of the dataset, select **Tabular** as the data type, choose **Regression/classification**, and then click **CREATE**.

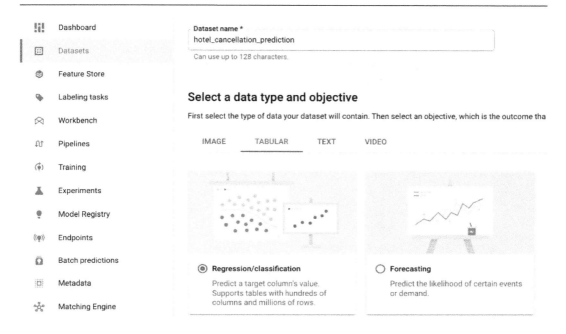

Figure 5.2 – Selecting a dataset type and model objective

3.  Upload the file named `hotel_reservation_data.csv` that you previously downloaded from the GitHub repository.

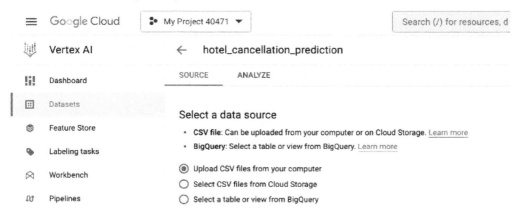

Figure 5.3 – Specifying a data source

4.  Enter a path to the GCS location where you would like to store the imported file. If you have not created a GCS bucket before, click on **Browse** and type in a name for the storage bucket you want to create. In subsequent prompts, pick the **Location** type as **Region and Region** for us-central1. For all other prompts, you can leave the default options already selected (*Figure 5.4*).

---

**Note**

It's important to have consistency in the location of the resources you create on GCP, so throughout this book, we will try to use us-central1 as the location.

---

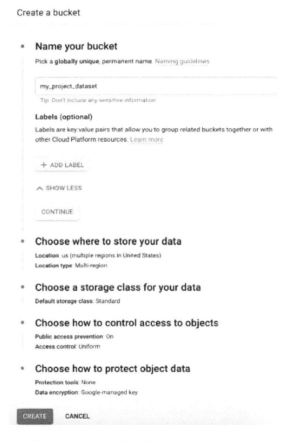

Figure 5.4 – Selecting cloud storage location

5.  Next, create a folder within the bucket where you want the imported file to be stored. Click the **Folder+** sign at the top right and then provide a name for the folder. Then, click **CREATE**. Finally, highlight the folder that was just created and click **Select**.

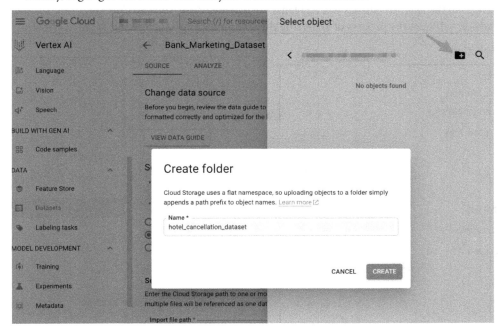

Figure 5.5 – Creating a folder to store the dataset

6.  Once the bucket and folder path have been selected, click **Continue** at the bottom of the screen, which will start the import of the CSV file into the Vertex AI dataset.

7.  Once the data import is completed, you will be taken to the **Analyze** tab. If you navigate away from the screen, you can always go back by following the **Vertex AI** > **Datasets** > `hotel_cancellation_prediction` or *<whatever name you specified for your dataset>* path. Here, all the feature statistics will be blank. To generate these, you can click **Generate Statistics**, which will start the process of analyzing the feature data and calculating detailed statistics.

Figure 5.6 – Generating statistics for the dataset

8.  Once this process is completed, you can click on a specific feature to see further details, such as the following:

    *   The distinct value count
    *   The missing percentage of data in the field
    *   The feature value distribution

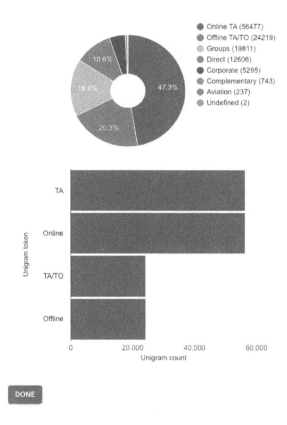

**Column name:** market_segment
**Missing % (count):** -
**Distinct values:** 8
**Most common value (%):** Online TA(47.305%)

Figure 5.7 – Analyzing key statistics for the dataset

The above screenshot shows the graphs explaining the distribution of feature/field titled 'market_ segment'.

Now let's look at how we will train an **AutoML Classification** model using the dataset discussed above.

## Training the AutoML model for tabular/structured data

Now, let's look at how you can use Vertex AI AutoML to train ML models on tabular data:

1.  Within the **ANALYZE** tab of the dataset that you want to use to train a model, click **TRAIN NEW MODEL | Other**.

Figure 5.8 – Training a new model

2.  Since we are trying to classify the reservations based on their cancellation likelihood, pick the model objective as **Classification**. Pick **AutoML** as the training method, which uses the codeless automated training option. Then, click **CONTINUE**.

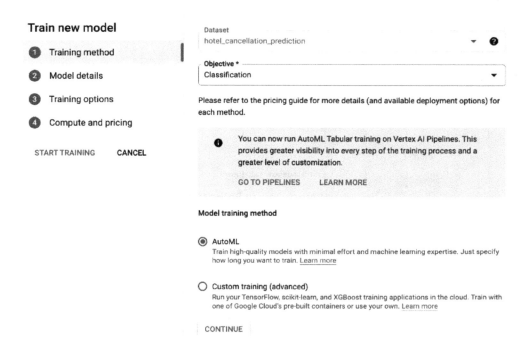

Figure 5.9 – Specifying the model type

3.  On the following **Model Details** screen, add these details:

    - Since it's the first time we are training a model on this dataset, pick **New Model**.

    - Type in what you want to call the model.

    - Select a target column so that AutoML knows which column to use as the prediction target. In our dataset, the column is titled `is_canceled`.

    - If you want the **Test** results to be exported to BigQuery for further analysis, select the **Export test dataset to big query** option, and provide a BigQuery table path where these results need to be stored. Vertex AI will create the table after the training run.

    - If you want Vertex AI to randomly split data into **Training**, **Validation**, and **Test** datasets, leave the default option, **Random**, selected. If you want to control the assignments of samples to the **Training**, **Validation**, and **Test** datasets, select the **Manual** option. In this case, you will need to provide a column where **Train/Validation/Test** assignments are provided.

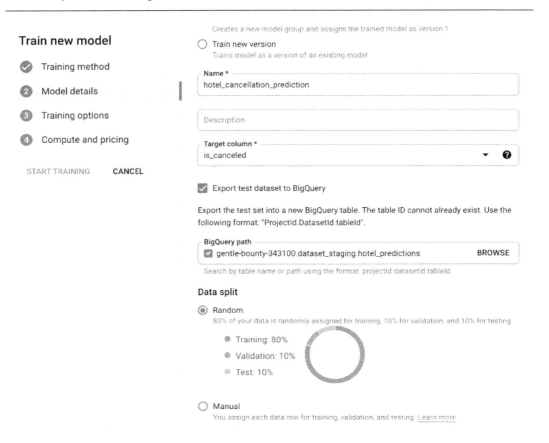

Figure 5.10 – Configuring the model training options

4.  On the following **Training Options** screen, verify that only the fields you want to be used for training are included in the current training run. Any unnecessary features should be removed, by selecting them and then clicking **Exclude from training** from the **Inclusivity** dropdown at the top.

    As shown in the following screenshot, remove the following two features from the training dataset we are using:

    *   `reservation_status`
    *   `reservation_status_date`

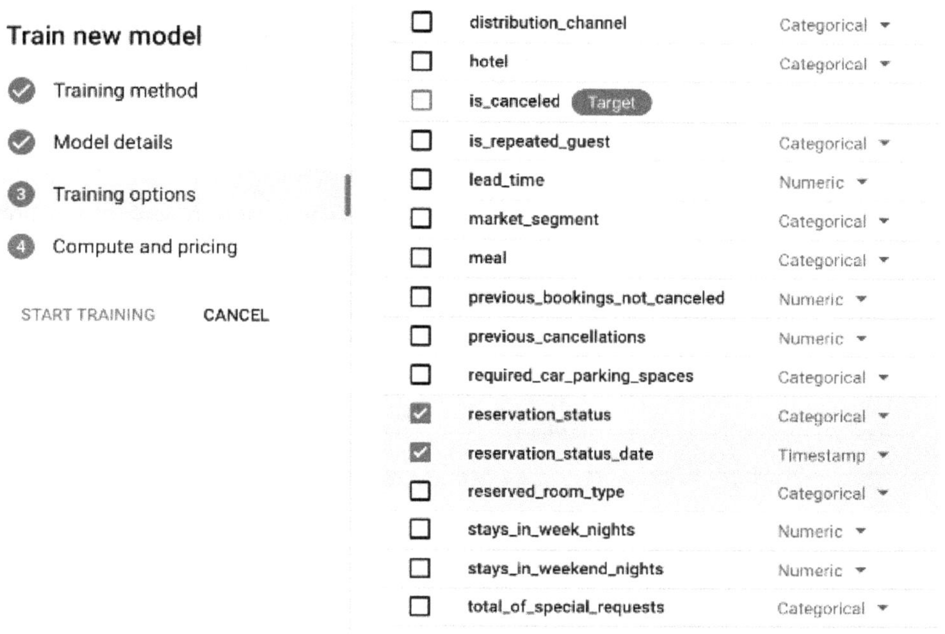

## Train new model

✓ Training method

✓ Model details

③ Training options

④ Compute and pricing

START TRAINING     CANCEL

| | | |
|---|---|---|
| ☐ | distribution_channel | Categorical ▾ |
| ☐ | hotel | Categorical ▾ |
| ☐ | is_canceled (Target) | |
| ☐ | is_repeated_guest | Categorical ▾ |
| ☐ | lead_time | Numeric ▾ |
| ☐ | market_segment | Categorical ▾ |
| ☐ | meal | Categorical ▾ |
| ☐ | previous_bookings_not_canceled | Numeric ▾ |
| ☐ | previous_cancellations | Numeric ▾ |
| ☐ | required_car_parking_spaces | Categorical ▾ |
| ☑ | reservation_status | Categorical ▾ |
| ☑ | reservation_status_date | Timestamp ▾ |
| ☐ | reserved_room_type | Categorical ▾ |
| ☐ | stays_in_week_nights | Numeric ▾ |
| ☐ | stays_in_weekend_nights | Numeric ▾ |
| ☐ | total_of_special_requests | Categorical ▾ |

Figure 5.11 – Selecting the features to be removed

Also, check that the Vertex AI has defaulted to the correct transformation types for each field. To be used for model training, tabular data must undergo a transformation process that is specific to each data feature. This transformation process indicates the function of a particular data feature. The supported types of transformations are as follows:

- **Categorical**: When training a model with a categorical feature in Vertex AI, the feature undergoes data transformations that help with the model training process. Vertex AI applies the following transformations to the feature and uses any that provide useful information:

  - The categorical string remains unchanged, with no modifications made to case, punctuation, spelling, tense, and other attributes.

  - The category name is converted into a dictionary lookup index, and an embedding is generated for each index.

  - Categories that appear less than five times in the training dataset are considered the **unknown** category. The **unknown** category is assigned a unique lookup index, and an embedding is generated for this category as well.

- **Text**: A feature that has undergone a text transformation is treated as freeform text and is usually made up of text tokens. The text is tokenized into words, and 1-grams and 2-grams are generated from those words. Each *n*-gram is then converted into a dictionary lookup index, and an embedding is generated for each index. Finally, the embeddings of all the elements are combined into a single embedding using the mean.

- **Numeric**: The following data transformations are applied to the feature, and any that provide useful information are used for training:

    - Rows with invalid numerical inputs, such as a string that cannot be parsed to `float32`, are not included in the training and prediction process.

    - The value is converted to `float32`.

    - The *z*-score of the value is calculated.

    - The value is bucketed based on quantiles, with a bucket size of 100.

    - The log of (`value+1`) is calculated when the value is greater than or equal to 0. If the value is less than 0, this transformation is not applied, and the value is considered a missing value.

    - The *z*-score of the log of (`value+1`) is calculated when the value is greater than or equal to 0. If the value is less than 0, this transformation is not applied, and the value is considered a missing value.

    - A Boolean value is assigned to indicate whether the value is `null`.

- **Timestamp**: The following data transformations are applied to the feature, and any that provide useful information are used for training:

    - The year, month, day, and weekday of the timestamp are determined and treated as categorical columns

    - Invalid numerical values, such as values outside the typical timestamp range or extreme values, are not removed or treated differently

    - The transformations for numerical columns are applied to the feature

    - Rows with invalid timestamp inputs, such as an invalid timestamp string, are not included in the training and prediction process

5. If you want to feed the model additional weights to rebalance the dataset, you can provide an additional column that contains weights assigned to each data sample.

6. As shown in the following screenshot, by default, the optimization objective is set to maximize the AUC ROC curve.

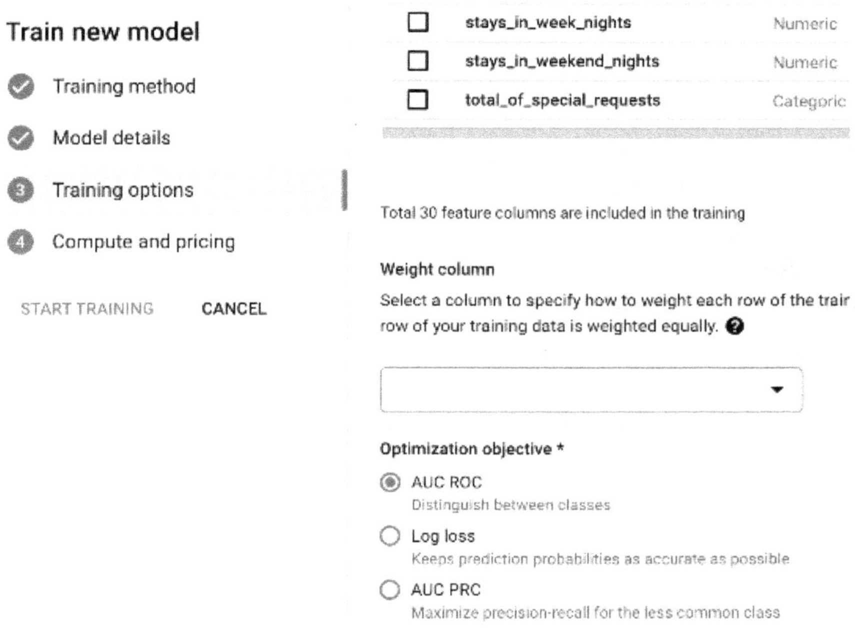

**Train new model**

✓ Training method

✓ Model details

3 Training options

4 Compute and pricing

START TRAINING     CANCEL

| | stays_in_week_nights | Numeric |
| | stays_in_weekend_nights | Numeric |
| | total_of_special_requests | Categoric |

Total 30 feature columns are included in the training

**Weight column**

Select a column to specify how to weight each row of the trair
row of your training data is weighted equally. ❓

▾

**Optimization objective ***

◉ AUC ROC
   Distinguish between classes

○ Log loss
   Keeps prediction probabilities as accurate as possible

○ AUC PRC
   Maximize precision-recall for the less common class

Figure 5.12 – Selecting the optimization objective

Although not required, you have the option to change the optimization objective.

For classification problems, there are five different objectives:

- **AUC ROC**: This objective in AutoML maximizes the area under the receiver operating characteristic curve. This is the default selection for binary classification and can be selected to distinguish between classes.

- **Log loss**: This aims to minimize the log loss for the model. It is used when the goal is to predict probabilities as accurately as possible, and it is the only supported objective in AutoML for multi-class classification models.

- **AUC PR**: This objective aims to maximize the area under the precision-recall curve. It optimizes results for predictions for the less common class.

- **Precision at recall**: This aims to maximize precision at a specific recall value.

- **Recall at precision**: This aims to maximize recall at a specific precision value.

For regression problems, there are three different objectives:

- **RMSE**: This objective aims to minimize the root mean squared error. It captures more extreme values accurately.

- **MAE**: This objective aims to minimize the mean absolute error. It views extreme values as outliers with less impact on the model.

- **RMSLE**: This objective aims to minimize the root mean squared log error. It penalizes errors in relative size rather than absolute value and is especially helpful when both predicted and actual values are quite large.

7.  Provide the maximum time you want the training to run. The minimum needs to be one hour. During the first run, it's hard to know how long the model will take to reach the highest possible accuracy, and you might have to experiment with this setting a little for new datasets and model types.

> **Best practice**
>
> For most small datasets (< 1 million data samples and < 20 features), two to four hours is a good starting point. If the **Enable early stopping** option is on, regardless of the number of hours you have budgeted for, the training will stop once AutoML determines that no further improvement in the model objective is being achieved with further rounds of training.

8.  Lastly, as shown in the following screenshot, click **START TRAINING** to kick off the training process. In a few hours, you will have a shiny new model ready to evaluate.

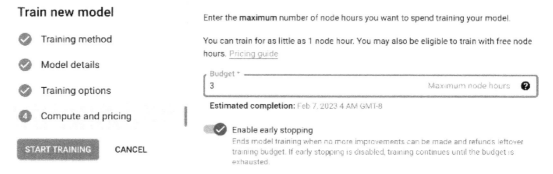

Figure 5.13 – Kicking off model training

Once the model training is completed, let's see how you can evaluate the model created by AutoML.

## Evaluating the trained model

Every time you train a machine learning model, it's crucial to evaluate its performance to determine whether it's reliable for real-world applications. Model evaluation metrics are calculated based on the model's performance against a portion of the dataset that was not used during training, referred to as the test dataset. This evaluation provides insight into how the model generalizes to new, unseen data and helps identify any issues or areas for improvement:

1. Go to **Model Registry** in the Vertex AI, and as shown in the following screenshot, locate the model you just trained. Click on the model, and then on the next screen, click the version of the model you just trained.

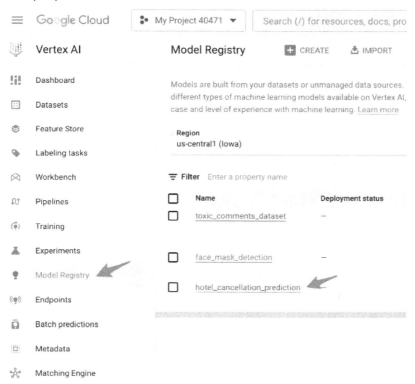

Figure 5.14 – Model Registry

2. On the **EVALUATE** tab, you will see the details of the test results generated by AutoML, by using the test dataset provided during training.

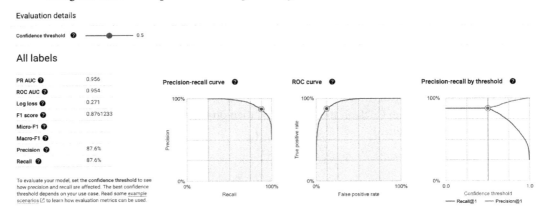

Figure 5.15 – The model evaluation metrics

Here is the key information typically shown in the Vertex AI **Evaluate** tab depending on the model type:

- Confidence threshold: When working with classification models, predictions are given a confidence score to indicate the level of certainty that the predicted class is correct. The score is a numeric assessment that determines how sure the model is that the prediction is accurate. For instance, consider a machine learning model that predicts whether a customer will cancel their hotel reservation:

  - To convert the score into a binary decision, a score threshold is used. The score threshold is the value at which the model says, "*Yes, the confidence score is high enough to conclude that the customer will cancel their reservation*" or "*No, the confidence score is not high enough to predict that the customer will cancel their reservation.*" The score threshold should be based on a specific use case, and a low score threshold increases the risk of misclassification.

- Precision and recall: In classification models, precision and recall are essential metrics to assess and summarize how well the model captures information and avoid errors:

  - Precision: Precision answers the question, "*Of all the predicted hotel reservation cancellations, how many were actually canceled?*" It measures the accuracy of the model's positive predictions – that is, the percentage of true positives out of all predicted positives.

  - Recall: Recall answers the question, "*Of all the canceled hotel reservations, how many did the model correctly predict?*" It measures the model's ability to identify true positive cases – that is, the percentage of true positives out of all actual positives.

Depending on the use case, you may need to optimize for either precision or recall. For instance, if you want to minimize the number of false negatives (i.e., hotel reservations that were canceled but were not identified by the model), then you should aim for a high recall. If you want to reduce the number of false positives (i.e., hotel reservations that were not canceled but were predicted to be), then you should aim for high precision.

Apart from precision and recall, there are several other classification metrics that are useful to evaluate the performance of a machine learning model. Here are some of the most commonly used metrics:

- **AUC PR**: The area under the **precision-recall** (**PR**) curve measures the trade-off between precision and recalls across various score thresholds. The **AUC PR** ranges from zero to one, where a higher value indicates a higher-quality model.

- **AUC ROC**: The area under the **receiver operating characteristic** (**ROC**) curve measures the model's performance across all possible score thresholds. The AUC ROC also ranges from zero to one, where a higher value indicates a higher-quality model.

- **Accuracy**: The fraction of classification predictions produced by the model that was correct. This is a simple and intuitive metric that provides an overall measure of the model's performance.

- **Log loss**: Log loss, also known as cross-entropy, serves as a crucial metric in assessing the alignment between a model's predictions and the actual target values. This metric quantifies the efficacy of the model's performance by measuring how closely its predictions match the real-world outcomes. With a scale spanning from zero to infinity, a lower log loss signifies a higher-quality model, showcasing its ability to make more accurate and confident predictions.

- **F1 score**: When seeking a balance between precision and recall, particularly in scenarios with imbalanced class distributions, the F1 score emerges as a valuable metric. This score represents the harmonic mean of precision and recall, operating on a scale ranging from zero to one. A higher F1 score denotes a model of superior quality, signifying its capability to achieve both precision and recall effectively, even in challenging class distribution scenarios.

By using these metrics, you can gain a more comprehensive understanding of your model's performance and make more informed decisions about how to improve it.

The **Evaluate** tab also showcases two additional useful pieces of information:

- **Confusion matrix**: A confusion matrix is a visualization tool that shows the frequency with which a model correctly predicts a result, and for the instances when it incorrectly predicts a result, the matrix shows what the model predicted instead. The confusion matrix is a helpful way to understand where the model may be "confusing" two results, and to diagnose the accuracy and performance of the model.

# Confusion matrix

This table shows how often the model classified each label correctly (in blue),

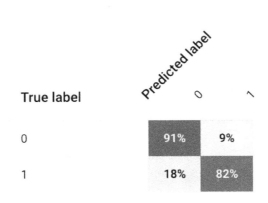

Figure 5.16 – The confusion matrix

- **Feature importance**: Model feature importance is expressed as a percentage for each feature, with a higher percentage indicating a stronger impact on model training. By looking at the feature importance values, we can gain a better understanding of the relative importance of different features in the model and how they contribute to the accuracy of predictions. The following screenshot shows the feature importance graph for the model we just trained.

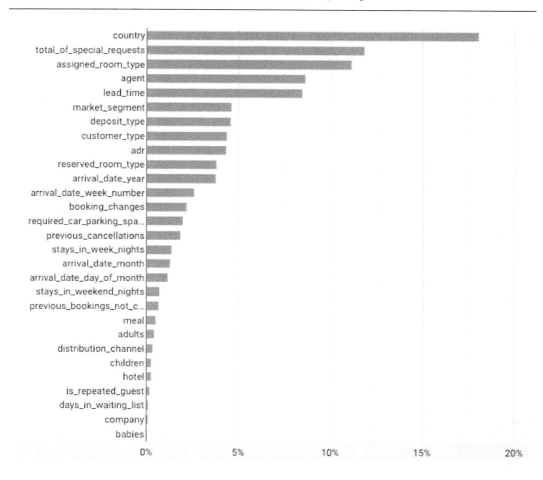

Figure 5.17 – Feature importance

So, we have trained a model and evaluated its key performance indicators. Now, let's look at how to deploy it and use it to generate predictions.

# Generating predictions using the recently trained model

Once the AutoML model is trained, you can generate predictions using one of the following two methods:

- **Batch predictions**: As the name suggests, batch predictions are asynchronous predictions generated for a batch of inputs. This is used when a real-time response is unnecessary and you want to submit a single request to process many data instances. In Vertex AI, a request for batch predictions can be submitted directly to a model residing in the Vertex AI Model registry, without the need to deploy it on an endpoint.

- **Online predictions**: If you need real-time inference – for example, when responding to application input – you need to use the Vertex AI online prediction option. To use online prediction, you must first deploy the model to an endpoint. This step provisions infrastructure resources and deploys prediction serving mechanism using the specified model, enabling it to serve predictions with low latency. The steps to deploy the model are shown in the following section.

> **Important note**
> Models deployed on Vertex AI endpoints continuously incur costs, regardless of their usage. This could add up quickly, depending on the type of underlying VM types, especially if you are using GPUs.

In the next section, we will deploy the ML model we trained on Vertex AI to generate online predictions.

# Deploying a model in Vertex AI

Now, let us walk you through the steps of deploying the trained model on Vertex AI to enable real-time predictions:

1. Go to **Model Registry**, click on the model and then the model version you want to deploy, and on the **DEPLOY & TEST** tab, click **DEPLOY TO ENDPOINT**.

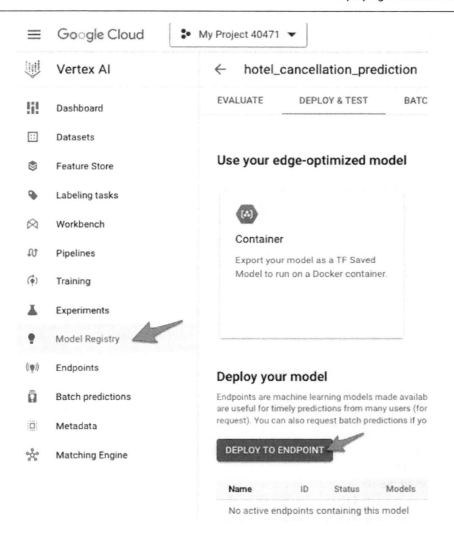

Figure 5.18 – Initiating model deployment

2.   Type in the desired name of the API endpoint being created and click **CONTINUE**.

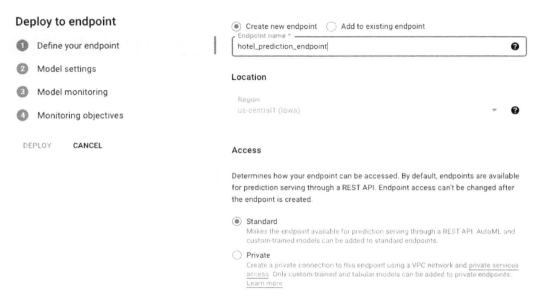

Figure 5.19 – Creating a model endpoint

3.   You can leave all default options unchanged for quick test deployment, but these are the settings you need to understand:

- **Traffic split**: If multiple versions of the model are deployed on the same API endpoint, this option allows users to define what percentage of total traffic is allocated to a specific version. For example, when deploying a new model, you might want only 2% of the overall incoming data to be routed to the new model so that it can be tested, while continuing to send the rest of the 98% of data to the existing version of the model. When deploying a model to a new endpoint, you need to leave this value at 100%, since there is no other model to split the workload.

- **Minimum number of compute nodes**: This is the bare minimum number of compute nodes always available to handle the inference requests. Even if there are no requests being handled, these nodes will constantly be deployed and incur charges.

- **Maximum number of compute nodes**: During autoscaling, as the number of incoming requests increases, Vertex AI will automatically increase the number of deployed nodes up to this number.

- **Machine type**: This relates to the configuration of the node supporting the online inference in terms of CPUs and memory.

- **Explainability options**: If you want the **Feature importance** value to be generated with every inference, select this option.

- Click **CONTINUE** once you have made the desired changes.

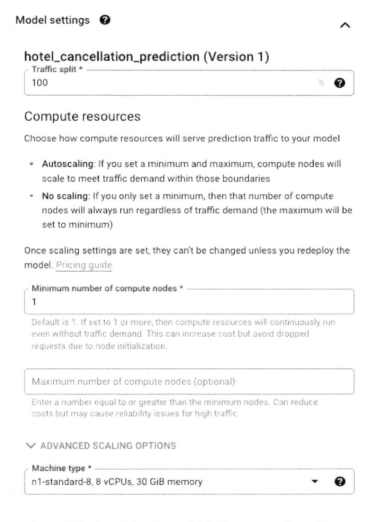

Figure 5.20 – Specifying the model deployment configurations

4.    On the **Model monitoring** page, turn off monitoring for now. We will cover model monitoring
      in detail in *Chapter 11, MLOps Governance with Vertex AI*. Click **DEPLOY**. This will start
      the endpoint creation process for new endpoints, followed by the model deployment process.

Once the model deployment is complete we can use the model to generate real time predictions as
shown below.

## Generating predictions

Once the model deployment is complete, you will see the endpoint listed in the **DEPLOY & TEST** tab.
Underneath that, there will be a **Test your model** table, listing all feature values required to generate
predictions. Fields will already have starting values based on the data used for AutoML training, but
you can type in different values and click **Predict** to generate predictions.

Figure 5.21 – Testing a deployed model

Once the model churns through the provided feature values, it will return the confidence score associated with each label. The one with the highest confidence score is the predicted label.

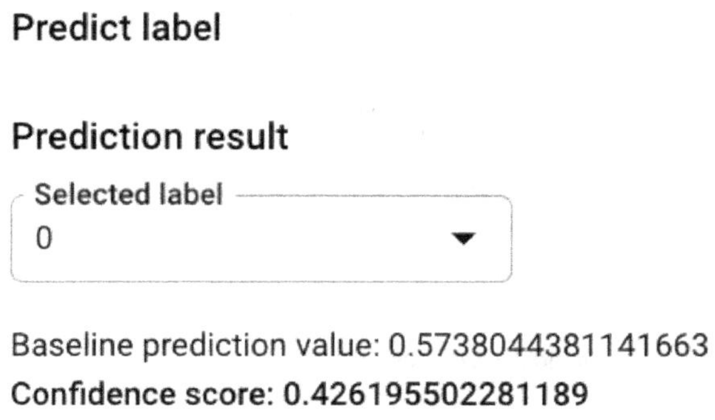

## Predict label

### Prediction result

┌─ Selected label ──────────────────────┐
│  0                                ▼   │
└────────────────────────────────────────┘

Baseline prediction value: 0.5738044381141663
**Confidence score: 0.426195502281189**

Figure 5.22 – The prediction result

Now let's look at the options available to developers to use the Vertex AI models programmatically.

# Generating predictions programmatically

To access the Vertex AI prediction service, you can work with the Vertex AI SDK for Python, or client libraries available for Python, Java, and Node.js. Check out *Install the Vertex AI client libraries* at `https://cloud.google.com/vertex-ai/docs/start/client-libraries` to learn how to install the client library for Java or Node.js. You can find a large number of Vertex AI sample notebooks, both community generate and those published by Google Cloud team at `https://github.com/GoogleCloudPlatform/vertex-ai-samples`.

If your preferred programming language doesn't have a client library, you can make use of the Vertex AI REST API instead.

## Submitting prediction requests using the REST API

Although the **DEPLOY & TEST** tab within Vertex AI makes it easy to test a model with a few samples, in most typical use cases, the model's API endpoint will be accessed programmatically by sending the input samples to the API and receiving a JSON response from it. Vertex AI makes it easier to get started by generating sample code, as shown in the following screenshot.

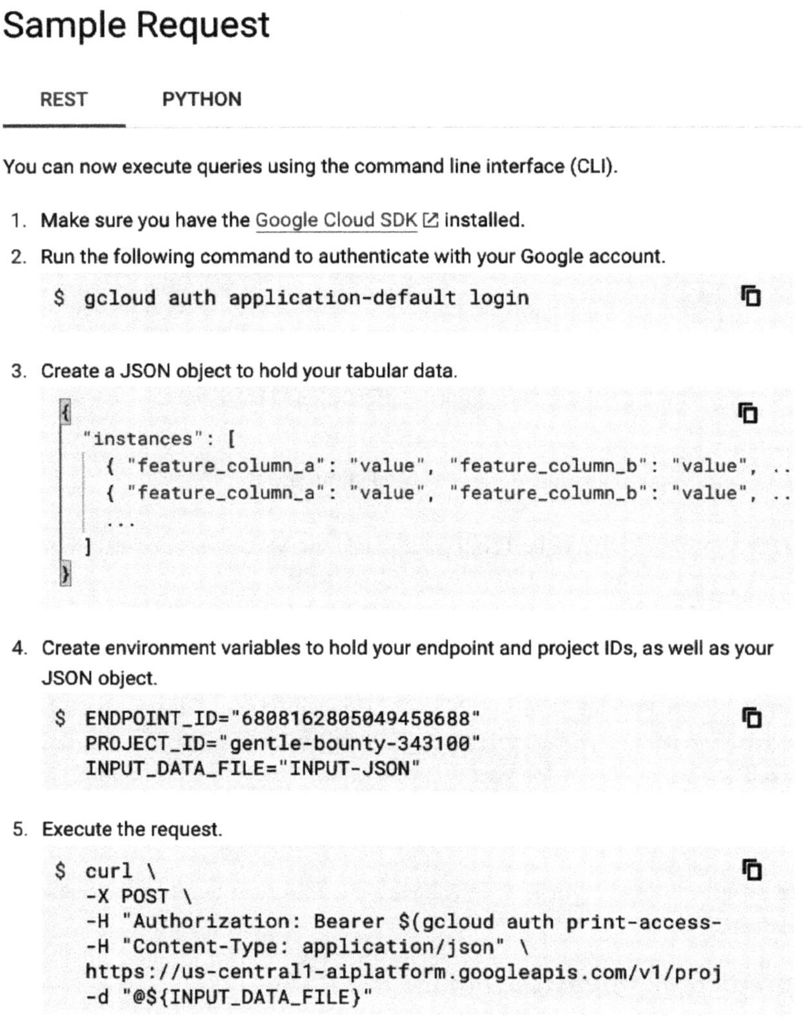

## Sample Request

**REST**    **PYTHON**

You can now execute queries using the command line interface (CLI).

1. Make sure you have the Google Cloud SDK ☑ installed.
2. Run the following command to authenticate with your Google account.

```
$ gcloud auth application-default login
```

3. Create a JSON object to hold your tabular data.

```
{
  "instances": [
    { "feature_column_a": "value", "feature_column_b": "value", ..
    { "feature_column_a": "value", "feature_column_b": "value", ..
    ...
  ]
}
```

4. Create environment variables to hold your endpoint and project IDs, as well as your JSON object.

```
$ ENDPOINT_ID="6808162805049458688"
  PROJECT_ID="gentle-bounty-343100"
  INPUT_DATA_FILE="INPUT-JSON"
```

5. Execute the request.

```
$ curl \
  -X POST \
  -H "Authorization: Bearer $(gcloud auth print-access-
  -H "Content-Type: application/json" \
  https://us-central1-aiplatform.googleapis.com/v1/proj
  -d "@${INPUT_DATA_FILE}"
```

Figure 5.23 – A sample REST request

We will use GCP's native Cloud Shell to submit a sample prediction request to the model deployed in Vertex AI. The first step is to open the cloud shell, as shown in the following screenshot.

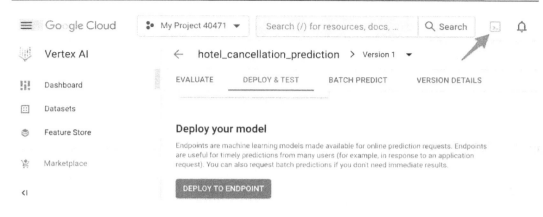

Figure 5.24 – Opening the cloud shell

Once Cloud Shell is open, upload the JSON file containing the input data samples to the Cloud Shell environment. We will submit this JSON payload as part of the request to the API.

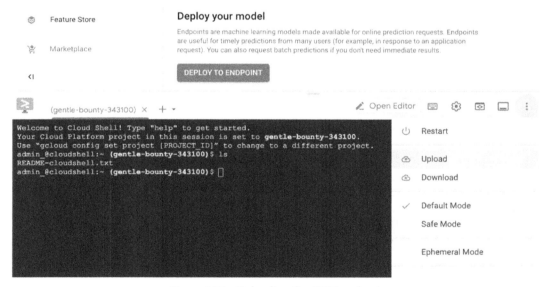

Figure 5.25 – Uploading the JSON payload

In Cloud Shell, copy and paste the following lines to set the environment variables after replacing `enpoint_id`, `project_id`, and the path to the input JSON file you uploaded.

Note that you need to replace `endpoint-id`, `project-id`, and the local path of the JSON file you just uploaded from your respective GCP environment:

```
ENDPOINT_ID="<your-endpoint-id"
PROJECT_ID="<your-project-id>"
INPUT_DATA_FILE="<path-to-input-json-file.json>"
```

Now, run the following command in Cloud Shell (you can copy and paste the command from the **Sample Request** screen we discussed previously:

```
curl \
-X POST \
-H "Authorization: Bearer $(gcloud auth print-access-token)" \
-H "Content-Type: application/json" \
https://us-central1-aiplatform.googleapis.com/v1/projects/${PROJECT_
ID}/locations/us-central1/endpoints/${ENDPOINT_ID}:predict \
-d "@${INPUT_DATA_FILE}"
```

Once the prediction request is sent to Vertex AI, it will run the data through the model hosted on the endpoint, and if everything goes well, you will receive a JSON response from the API.

Here's the response received from the Vertex AI endpoint:

```
"predictions": [
  {"scores": [
      0.8947222328186035,
      0.1052777245640755],
    "classes": ["0","1"]},
  {"classes": ["0","1"],
    "scores": [
      0.8825596570968627 9,
      0.117440328001976]}],
    "deployedModelId": "3558930023110934528",
    "model": "projects/217932574390/locations/us-central1/
models/6845069012347387904",
    "modelDisplayName": "hotel_cancellation_prediction,"
    "modelVersionId": "1"}
```

The response contains three main parts – `predictions`, `deployedModelId`, and `model`.

The `predictions` section contains the actual predictions made by the model. In this case, since we submitted two samples for prediction, two predictions have been made, each with a set of scores and predicted classes. The scores represent the model's confidence in each class, and the classes represent the label or category that the model has predicted for each prediction.

- The first prediction has scores of 0.8947 for class `"0"` and 0.1053 for class `"1"`, which suggests that the model is highly confident that the first prediction belongs to class `"0"`

- The second prediction has scores of 0.8826 for class `"0"` and 0.1174 for class `"1"`, which also suggests that the model is confident that the second prediction belongs to class `"0"` but is less certain than it was about the first prediction

`deployedModelId` is a unique identifier for the deployed model. This is useful to keep track of different versions of the same model that may be deployed at different times, or in different locations.

The `model` section provides information about the model that was used to make the predictions. It includes the model's ID, location, and display name, as well as the version of the model that was used for these predictions (in this case, version 1). This information can help debug and troubleshoot if there are issues with the model or its predictions.

In the preceding end-to-end example, we walked you through building an AutoML classification model for tabular/structured data. In the subsequent chapters, we will also provide examples of how to build AutoML Vision models using image data (*Chapter 16*) and examples of AutoML NLP models with text data (*Chapter 17*).

# Summary

In this chapter, we discussed the transformative influence of no-code platforms in democratizing AI and machine learning, allowing individuals without specialized coding knowledge to develop sophisticated models. You also learned about the no-code machine learning options available through Google Cloud's Vertex AI toolset, and we explored the steps to build, evaluate, deploy, and generate predictions with an AutoML model. The chapter serves as a practical guide for users interested in leveraging Google Cloud's machine learning capabilities without writing code.

In the next chapter, we will explore some low-code options to build a machine learning model. Although that would require some additional technical skills compared to the no-code options discussed in this book, it would also give you more fine-grained control over how your model is built.

# 6

# Low-Code Options
# for Building ML Models

**BigQuery Machine Learning**, often abbreviated as **BQML**, is a tool offered as part of Google Cloud that seamlessly merges the worlds of data warehousing and ML. Designed to bridge the gap between data analysts and ML models, BQML empowers individuals to build, evaluate, and predict with ML models directly within the confines of BigQuery, without the need to move data or master a new toolset.

This integration not only simplifies the process of model creation but also presents an intuitive transition for those familiar with SQL. With just a few statements, you can go from data analysis to predictive insights.

In this chapter, we will go over the following topics:

- What is BQML?
- Using BQML for feature transformations
- Building ML models with BQML
- Doing inference with BQML

## What is BQML?

BQML is a powerful, built-in ML service provided by Google Cloud that allows users to create, train, and deploy ML models using familiar SQL queries. BQML is designed to simplify the process of building and deploying ML models for those who may not have a strong background in data science or programming. In this chapter, we will explore the key features and capabilities of BQML and how you can use it to leverage the power of Google Cloud AI for your projects.

BQML provides a seamless way to integrate ML into your data analytics workflows without requiring a deep understanding of ML concepts or programming languages. With BQML, you can do the following:

- Create and train ML models using SQL queries
- Make predictions using trained models
- Evaluate the performance of your models
- Perform feature transformation and hyperparameter tuning
- Understand model explanations and weights
- Export and import models

Utilizing BQML offers numerous advantages:

- BQML eliminates the need to load data into local memory, thereby addressing the constraint posed by sizable datasets
- BQML streamlines the ML process by handling standard tasks such as dividing data into train and test sets, selecting and adjusting learning rates, and choosing an optimization approach
- With automatic versioning of BQML models, tracking alterations and reverting to earlier versions when needed becomes effortless
- When serving predictions, BQML models can be seamlessly integrated with Vertex AI

There are also a few limitations to using BQML:

- **Limited model types**: BQML supports a restricted set of ML models, such as linear regression, logistic regression, k-means clustering, matrix factorization, and others. It may not meet the requirements of projects that necessitate advanced or specialized models.
- **Customizability**: BQML's automated approach to ML means that there is limited scope for customization. Users might not be able to fine-tune models or experiment with different model architectures as they could with other ML frameworks.
- **Scalability**: Although BQML is designed for handling large datasets, it may not scale as effectively as other distributed ML frameworks when working with extremely large datasets or complex models.
- **Limited support for deep learning**: BQML does not support training for deep learning models for unstructured data, which may be essential for certain use cases such as image recognition, natural language processing, or speech recognition. However, it does provide options to import pre-trained TensorFlow, TensorFlow Lite, and ONNX models trained on image data. Once imported into BigQuery, you can use the `ML.PREDICT` function to make predictions based on images. BQML now also supports adding **remote models** as API endpoints, which opens up the possibility of adding any model hosted on Vertex AI endpoints or adding other cloud-based ML services such as Vision API to support additional use cases.

- **Feature engineering**: BQML might not be the best option for extensive feature engineering as it focuses more on simplifying the ML process. Users may need to perform feature engineering outside of BQML for advanced feature engineering tasks. We will discuss the limitations in more detail in the *Feature engineering* section of this chapter.

- **External data sources**: BQML works primarily with Google BigQuery data, limiting its flexibility in terms of data sources. If you want to use data from different sources or formats, you might need to import it into BigQuery first.

- **Model portability**: BQML models are tightly integrated with Google Cloud. Exporting models for use outside of the Google ecosystem may be challenging and might require additional work.

Now let's look at how you can start using BigQuery for ML solutions.

## Getting started with BigQuery

Google BigQuery is a serverless, fully managed data warehouse that enables super-fast SQL queries using the processing power of Google's infrastructure. Since BigQuery is not part of Vertex AI, we won't be covering the features of the tool in depth in this book, but here's a quick guide on getting started. This should be enough information to help you follow along with the exercises later in this chapter:

1. **Set up a Google Cloud project**: Before you can use BigQuery, you'll need to set up a **Google Cloud Platform** (**GCP**) project. Head over to the Google Cloud console and create a new project. If you've never used GCP before, you might need to create an account and set up billing information.

2. **Enable the BigQuery API**: Within your GCP project, navigate to the **API & Services** section and enable the BigQuery API.

3. **Access the BigQuery console**: Once the API has been enabled, you can access the BigQuery console either via the GCP dashboard or directly through the BigQuery console link (`https://console.cloud.google.com/bigquery`).

4. **Create a dataset**: Datasets are containers for tables, views, and other data objects in BigQuery. To create one, click on the vertical ellipsis next to your GCP project name in the BigQuery console, select **Create Dataset**, and fill in the dataset's name. Then, click **Create Dataset**.

5. **Load data**: BigQuery supports various data formats, including CSV, JSON, and others. You can load data into BigQuery from Google Cloud Storage, send data directly with an API request, or manually upload files. To load data, navigate to your dataset in the BigQuery console on the left, click **Create Table**, and follow the prompts.

6. **Run SQL queries**: With data loaded into BigQuery, you can then run SQL queries. Use the query editor in the BigQuery console to start analyzing your data using SQL.

Now, let's look at how you can use BigQuery's native functions to do large-scale feature/data transformations to prepare training data for the ML models.

## Using BQML for feature transformations

Two types of feature preprocessing are supported by BQML:

- **Automatic preprocessing**: During training, BQML carries out automatic preprocessing. For further details, please carries out automatic preprocessing like missing data imputation, one-hot encoding, and timestamp transformation and encoding.

- **Manual preprocessing**: You can use the TRANSFORM clause provided by BQML to define customized preprocessing using manual preprocessing functions. These functions can also be utilized outside the TRANSFORM clause.

While BQML does support some feature engineering tasks, it has certain limitations compared to more flexible and feature-rich ML frameworks:

- **Limited preprocessing functions**: BQML provides a basic set of SQL functions for data preprocessing, such as scaling and encoding. However, it may lack some advanced preprocessing techniques or specialized functions available in other ML libraries such as **scikit-learn** or **TensorFlow**.

- **No automated feature selection**: BQML does not offer automated feature selection methods to identify the most important variables in your dataset. You must manually select and engineer features based on your domain knowledge and intuition, or use external tools for feature selection.

- **Complex feature transformations**: BQML's SQL-based approach may not be well suited for certain complex feature transformations that involve non-linear combinations, rolling windows, or sequential patterns in the data. In such cases, you may need to preprocess your data using other tools or programming languages before using BQML.

- **Custom feature generation**: BQML lacks the flexibility to create custom features, such as domain-specific functions or transformations, as easily as you can with more versatile ML libraries. You might need to implement these custom features outside of BQML, which could be cumbersome and less efficient.

- **Feature engineering pipelines**: BQML does not provide a built-in mechanism to create and manage reusable feature engineering pipelines. In contrast, other ML frameworks offer functionality to build modular and maintainable pipelines, streamlining the process of applying the same transformations to training and validation datasets or during model deployment.

While BQML simplifies the ML process, it may not be the best choice for projects that require extensive or advanced feature engineering. In such cases, you may need to preprocess your data using external tools or libraries and then import the transformed data into BigQuery for further analysis with BQML.

# Manual preprocessing

BQML provides a variety of manual preprocessing functions that can be utilized with the CREATE MODEL syntax to preprocess your data before training. These functions can also be used outside the TRANSFORM clause. These preprocessing functions can be scalar, operating on a single row, or analytic, operating on all rows and outputting results based on statistics collected across all rows. When ML analytic functions are used inside the TRANSFORM clause during training, the same statistics are automatically applied to the input in prediction.

The following table lists all the supported data preprocessing functions in BigQuery:

| Function Name | Description |
| --- | --- |
| ML.BUCKETIZE | Bucketizes a numerical expression into user-defined categories based on provided split points |
| ML.POLYNOMIAL_EXPAND | Generates polynomial combinations of a given set of numerical features up to a specified degree |
| ML.FEATURE_CROSS | Generates feature crosses of categorical features up to a specified degree |
| ML.NGRAMS | Extracts n-grams from an array of tokens, based on a given range of $n$ values |
| ML.QUANTILE_BUCKETIZE | Bucketizes a numerical expression into quantile-based categories based on several buckets |
| ML.HASH_BUCKETIZE | Bucketizes a string expression into a fixed number of hash-based buckets |
| ML.MIN_MAX_SCALER | Scales a numerical expression to the range [0, 1] capped with MIN and MAX across all rows |
| ML.STANDARD_SCALER | Standardizes a numerical expression |
| ML.MAX_ABS_SCALER | Scales a numerical expression to the range [-1, 1] by dividing through the largest maximum absolute value |
| ML.ROBUST_SCALER | Scales a numerical expression using statistics that are robust to outliers |
| ML.NORMALIZER | Normalizes an array expression to have a unit norm using the given p-norm |
| ML.IMPUTER | Replaces NULL in an expression using a specified value (for example, mean, median, or most frequent) |
| ML.ONE_HOT_ENCODER | Encodes a string expression using a one-hot encoding scheme |
| ML.LABEL_ENCODER | Encodes a string expression to an INT64 in [0, n_categories] |

Table 6.1 – Data transformation functions

Here is a list of all the functions with their inputs and outputs, as well as an example for each:

- `ML.BUCKETIZE`

  Bucketizes a numerical expression into user-defined categories based on provided split points.

  Input:

  - `numerical_expression`: Numerical expression to bucketize.

  - `array_split_points`: Sorted numerical array with split points.

  - `exclude_boundaries` (optional): If `TRUE`, the two boundaries are removed from `array_split_points`. The default value is `FALSE`.

  Output:

  `STRING` is the name of the buckets into which the `numerical_expression` field is split.

  Here's an example SQL statement:

  ```
  WITH dataset AS (
  SELECT 21 AS age UNION ALL
  SELECT 33 AS age UNION ALL
  SELECT 45 AS age UNION ALL SELECT 59 AS age UNION ALL
  SELECT 66 AS age )
  SELECT age, ML.BUCKETIZE(age, [18, 35, 50, 65]) AS age_bucket
  FROM dataset;
  ```

  Submitting this query in BigQuery should generate the following output:

  ```
  age   age_bucket
  21    bin_2
  33    bin_2
  45    bin_3
  59    bin_4
  bin_5
  ```

- `ML.FEATURE_CROSS`

  This generates feature crosses of categorical features up to a specified degree:

  ```
  Input: array_expression, degree
  Output: ARRAY<STRUCT<name STRING, value FLOAT64>>
  ```

  Here's an example SQL statement:

  ```
  WITH dataset AS (
  SELECT "dog" AS animal, "brown" AS color UNION ALL
  SELECT "cat" AS animal, "black" AS color UNION ALL
  SELECT "bird" AS animal, "yellow" AS color UNION ALL
  SELECT "fish" AS animal, "orange" AS color)
  ```

```
SELECT animal, color, ML.FEATURE_CROSS(STRUCT(animal, color)) AS
animal_color
FROM dataset;
```

Submitting this query in BigQuery should generate the following output:

| animal | color | animal_color |
|--------|-------|--------------|
| dog | brown | dog_brown |
| cat | black | cat_black |
| bird | yellow | bird_yellow |
| fish | orange | fish_orange |

Table 6.2 – BigQuery Query Output

Note that the `ML.FEATURE_CROSS` function can be used to create a cross of multiple columns if you include more columns in the `ARRAY` argument.

• `ML.NGRAMS`

This extracts n-grams from an array of tokens, based on a given range of *n* values.

Input:

• `array_input`: `ARRAY` of `STRING`. The strings are the tokens to be merged.

• `range`: `ARRAY` of two `INT64` elements or a single `INT64`. These two sorted `INT64` elements in the `ARRAY` input are the range of n-gram sizes to return. A single `INT64` is equivalent to the range of [x, x].

• `separator`: Optional `STRING`. `separator` connects two adjacent tokens in the output. The default value is whitespace.

Output: `ARRAY` of `STRING`.

Here's an example SQL statement:

```
WITH dataset AS (
SELECT "apple" AS fruit, "cherry" AS fruit2,"pear" AS fruit3
UNION ALL
SELECT "banana" AS fruit, "banana" AS fruit2,"melon" AS fruit3
UNION ALL
SELECT "cherry" AS fruit, "cherry" AS fruit2, "pineapple" AS
fruit3)
SELECT fruit,fruit2,fruit3, ML.NGRAMS([fruit,fruit2,fruit3],
[2]) AS fruit_ngrams
FROM dataset;
```

Submitting this query in BigQuery should generate the following output:

| fruit | fruit2 | fruit3 | fruit_ngrams |
|---|---|---|---|
| apple | cherry | pear | [apple cherry, cherry pear] |
| banana | banana | melon | [banana banana, banana melon] |
| cherry | cherry | pineapple | [cherry cherry, cherry pineapple] |

Table 6.3 – Output from BigQuery

- ML.QUANTILE_BUCKETIZE

This bucketizes a numerical expression into quantile-based categories based on several buckets.

Input:

- numerical_expression: Numerical expression to bucketize

- num_buckets: INT64. The number of buckets to split numerical_expression into

Output: STRING.

Here's an example SQL statement:

```
WITH dataset AS (
SELECT 21 AS age UNION ALL
SELECT 33 AS age UNION ALL
SELECT 45 AS age UNION ALL
SELECT 59 AS age UNION ALL
SELECT 66 AS age)
SELECT age, ML.QUANTILE_BUCKETIZE(age, 4) OVER() AS age_bucket
FROM dataset
ORDER BY age;
```

In this example, we create a virtual table dataset with five rows of age data. Then, we use the ML.QUANTILE_BUCKETIZE function to bucketize the age column into four quantile buckets. The resulting age_bucket column shows which quantile bucket each row of the dataset belongs to.

Here's the output:

```
age  age_bucket
21   bin_1
33   bin_2
45   bin_3
59   bin_4
66   bin_4
```

- `ML.HASH_BUCKETIZE`

  This bucketizes a string expression into a fixed number of hash-based buckets.

  Input:

  - `string_expression`: `STRING`. The string expression to bucketize.

  - `hash_bucket_size`: `INT64`. The number of buckets. Expected `hash_bucket_size >= 0`. If `hash_bucket_size = 0`, the function only hashes the string without bucketizing the hashed value.

  Output: `INT64`.

  Here's an example SQL statement:

  ```
  WITH dataset AS (
  SELECT "horse" AS animal UNION ALL
  SELECT "cat" AS animal UNION ALL
  SELECT "dog" AS animal UNION ALL
  SELECT "fish" AS animal)
  SELECT animal, ML.HASH_BUCKETIZE(animal, 2) AS animal_bucket
  FROM dataset;
  ```

  In this example, we create a virtual table dataset with four rows of data. Then, we use the `ML.HASH_BUCKETIZE` function to hash the `animal` column into two buckets. The resulting `animal_bucket` column shows which hash bucket each row of the dataset belongs to.

  Note that the `ML.HASH_BUCKETIZE` function can be used to hash the values of a column into a different number of buckets by specifying a different value for the second argument.

  Here's the output:

  ```
  animal   animal_bucket
  horse    1
  cat      0
  dog      0
  fish     0
  ```

  In this example, we create a virtual table dataset with four rows of data. Then, we use the `ML.HASH_BUCKETIZE` function to hash the animal column into two buckets. The resulting `animal_bucket` column shows which hash bucket each row of the dataset belongs to.

- `ML.MIN_MAX_SCALER`

  Scales a numerical expression to the range [0, 1] capped with `MIN` and `MAX` across all rows.

  Input: `numerical_expression`. Numerical expression to scale.

  Output: `DOUBLE`.

Here's an example SQL statement:

```
WITH dataset AS (
SELECT 10 AS age UNION ALL
SELECT 20 AS age UNION ALL
SELECT 30 AS age UNION ALL
SELECT 40 AS age UNION ALL
SELECT 50 AS age)
SELECT age, ML.MIN_MAX_SCALER(age) Over() AS scaled_age
FROM dataset;
```

In this example, we create a virtual table dataset with five rows of age data. Then, we use the `ML.MIN_MAX_SCALER` function to scale the `age` column to a range of 0 to 1. Note that the `ML.MIN_MAX_SCALER` function can be used to scale the values of a column to a different range by specifying different values for the `MIN` and `MAX` arguments.

Here's the output:

```
age   scaled_age
50    1.0
20    0.25
40    0.75
10    0.0
30    0.5
```

- `ML.STANDARD_SCALER`

  This function standardizes a numerical expression.

  Input: `numerical_expression`. Numerical expression to scale.

  Output: DOUBLE.

  Here's an example SQL statement:

```
WITH dataset AS (
SELECT 10 AS age UNION ALL
SELECT 20 AS age UNION ALL
SELECT 30 AS age UNION ALL
SELECT 40 AS age UNION ALL
SELECT 50 AS age)
SELECT age, ML.STANDARD_SCALER(age) OVER() AS scaled_age
FROM dataset;
```

In this example, we create a virtual table dataset with five rows of age data. Then, we use the `ML.STANDARD_SCALER` function to standardize the `age` column to have a mean of 0 and a standard deviation of 1. The resulting `scaled_age` column shows the standardized values of the `age` column.

Here's the output:

```
age   scaled_age
40    0.63245553203367588
10    -1.2649110640673518
50    1.2649110640673518
20    -0.63245553203367588
30    0.0
```

- ML.MAX_ABS_SCALER

  This function scales a numerical expression to the range [-1, 1] by dividing through the largest maximum absolute value.

  Input: numerical_expression.

  Output: DOUBLE.

  Here's an example SQL statement:

  ```
  WITH dataset AS (
  SELECT -10 AS age UNION ALL
  SELECT 20 AS age UNION ALL
  SELECT -30 AS age UNION ALL
  SELECT 40 AS age UNION ALL
  SELECT -50 AS age)
  SELECT age, ML.MAX_ABS_SCALER(age)
  OVER() AS scaled_age
  FROM dataset;
  ```

  In this example, we create a virtual table dataset with five rows of age data. Then, we use the ML.MAX_ABS_SCALER function to scale the age column so that the absolute values of the largest magnitude element in the column are scaled to 1. The resulting scaled_age column shows the scaled values of the age column.

  Here's the output:

  ```
  age   scaled_age
  -10   -0.2
  -50   -1.0
  20    0.4
  40    0.8
  -30   -0.6
  ```

- ML.NORMALIZER

  This function normalizes array_expression to have a unit norm using the given p-norm.

  Input: array_expression, p.

  Output: ARRAY<DOUBLE>.

Here's an example SQL statement:

```
WITH dataset AS (
SELECT 1 AS x, 2 AS y UNION ALL
SELECT 3 AS x, 4 AS y UNION ALL
SELECT 5 AS x, 6 AS y UNION ALL
SELECT 7 AS x, 8 AS y UNION ALL
SELECT 9 AS x, 10 AS y)
SELECT x, y, ML.NORMALIZER([x, y], 1) AS norm_xy
FROM dataset;
```

Here's the output:

```
x   y   norm_xy
1   2   "[0.447213595499579,0.8944271909999158]"
3   4   "[0.6,0.8]"
5   6   "[0.6401843996644799,0.7682212795973758]"
7   8   "[0.658504607868518,0.75257669470687782]"
9   10   "[0.6689647316224497,0.7432941462471663]"
```

- ML.IMPUTER

  This function replaces NULL in an expression using a specified value (for example, mean, median, or most frequent).

  Input: expression, strategy.

  Output: DOUBLE for numerical expression. STRING for STRING expression.

  Here's an example SQL statement:

```
WITH dataset AS (
SELECT 10 AS age, 20 AS height UNION ALL
SELECT 20 AS age, 30 AS height UNION ALL
SELECT 30 AS age, 40 AS height UNION ALL
SELECT 40 AS age, 50 AS height UNION ALL
SELECT 50 AS age, NULL AS height)
SELECT age, height, ML.IMPUTER(height,"median") OVER() AS
imputed_height FROM dataset;
```

  Here's the output:

```
age   height   imputed_height
20    30       30.0
10    20       20.0
40    50       50.0
50    null      30.0
30    40       40.0
```

- `ML.ONE_HOT_ENCODER`

This function encodes `string_expression` using a one-hot encoding scheme.

Inputs:

- `string_expression`: The `STRING` expression to be encoded.

- `drop` (optional): This determines which category to drop during encoding. The default value is `none`, which means all categories are retained.

- `top_k` (optional): `INT64`. This limits the encoding vocabulary to the `top_k` frequent categories. The default value is 32,000, and the max supported value is 1 million to avoid suffering from high dimensionality.

- `frequency_threshold` (optional): `INT64`. It limits the encoding vocabulary to categories whose frequency is `>= frequency_threshold`. The default value is 5.

Output: It is an array of `STRUCT` that contains the encoded values, where `index` is the index of the encoded value and `value` is the value of the encoded value.

Here's an example SQL statement:

```
WITH
  input_data AS (
    SELECT 'red' AS color UNION ALL
    SELECT 'blue' AS color UNION ALL
    SELECT 'green' AS color UNION ALL
    SELECT 'green' AS color UNION ALL
    SELECT 'purple' AS color),
  vocab AS (
    SELECT color, COUNT(*) AS frequency
    FROM input_data
    GROUP BY color)
SELECT color,
  ML.ONE_HOT_ENCODER(color) OVER()  AS encoding
FROM input_data
```

The preceding query results in the following output:

| color | encoding.index | encoding.value |
| --- | --- | --- |
| green | 0 | 1.0 |
| red | 0 | 1.0 |
| purple | 0 | 1.0 |
| blue | 0 | 1.0 |
| green | 0 | 1.0 |

Table 6.4: Output from the previous query

- `ML.LABEL_ENCODER`

This function converts string values into `INT64` numbers within a designated range. The function organizes the encoding terms in alphabetical order, and any category not found in this vocabulary will be represented as `0`. When utilized in the `TRANSFORM` clause, the vocabulary and categories omitted during the training process are seamlessly applied during prediction.

Inputs:

- `string_expression`: The `STRING` expression to be encoded.

- `top_k`: Optional `INT64`. This limits the encoding vocabulary to the `top_k` frequent categories. The default value is 32,000, and the max supported value is 1 million.

- `frequency_threshold`: Optional `INT64`. This limits the encoding vocabulary to categories whose frequency is >= `frequency_threshold`. The default value is 5.

Output: `INT64`. This is the encoded value of the string expression in the specified range.

Here's an example SQL statement:

```
WITH data AS (
  SELECT 'apple' AS fruit UNION ALL
  SELECT 'banana' UNION ALL
  SELECT 'orange' UNION ALL
  SELECT 'apple' UNION ALL
  SELECT 'pear' UNION ALL
  SELECT 'kiwi' UNION ALL
  SELECT 'banana')
SELECT fruit, ML.LABEL_ENCODER(fruit, 2,2) OVER() AS encoded_
fruit FROM data
```

The preceding query results in the following output:

| fruit | encoded_fruit |
|--------|---------------|
| orange | 0 |
| pear | 0 |
| banana | 2 |
| apple | 1 |
| kiwi | 0 |
| apple | 1 |
| banana | 2 |

Table 6.5: Output from the previous query

Now let's look at the different type of ML models you can build with BQML.

# Building ML models with BQML

BQML supports model training for several different use cases. The key model categories that are currently supported are supervised learning models, unsupervised learning models, time series models, imported models, and remote models.

The following table showcases some of the key ML model types that are supported within BigQuery:

| Model Type | Model Types | Manually Defined Feature Preprocessing | Hyperparameter Tuning in BQML |
| --- | --- | --- | --- |
| Supervised | Linear and logistic regression | Supported | Supported |
| Supervised | Deep neural networks | Supported | Supported |
| Supervised | Wide-and-deep | Supported | Supported |
| Supervised | Boosted trees | Supported | Supported |
| Supervised | Random forest | Supported | Supported |
| Supervised | AutoML tables | Not supported | Automated |
| Unsupervised | k-means | Supported | Supported |
| Unsupervised | Matrix factorization | Not supported | Supported |
| Unsupervised | PCA | Supported | Not supported |
| Unsupervised | Autoencoder | Supported | Not supported |
| Time series | `ARIMA_PLUS` | Only automatic preprocessing | Supported (`auto.ARIMA4`) * |
| Time series | `ARIMA_PLUS_XREG` | Only automatic preprocessing | Supported (`auto.ARIMA4`) * |

Table 6.6 – Supported capabilities for key ML models

There are two other important model creation options available in BigQuery to help you utilize ML models built outside BigQuery – imported models and remote models.

BQML allows you to import models that have been trained outside BigQuery so that they can be used for inference within BigQuery.

The following model frameworks are supported for importing:

- TensorFlow
- TensorFlow Lite
- ONNX
- XGBoost

BQML allows you to register existing Vertex AI endpoints as a remote model. Once registered in BigQuery, you can send a prediction request to the Vertex AI endpoint from within BigQuery for inference.

# Creating BQML models

The BigQuery function that's used to initiate model creation is aptly called CREATE. In this section, we'll look at the options available to a user when they're creating different types of BQML models using the CREATE function. You don't necessarily need to read through the details of every single model at the moment. This should be used more as a reference, as needed.

## *Linear or logistic regression models*

The following is the syntax for creating regression models, along with the different required and optional arguments you need to provide as part of the query:

```
{CREATE OR REPLACE MODEL} model_name
[OPTIONS(MODEL_TYPE = { 'LINEAR_REG' | 'LOGISTIC_REG' },
    INPUT_LABEL_COLS = string_array,
    OPTIMIZE_STRATEGY = { 'AUTO_STRATEGY'  },
    L1_REG = float64_value,
    L2_REG = float64_value,
    MAX_ITERATIONS = int64_value,
    LEARN_RATE_STRATEGY = { 'LINE_SEARCH' | 'CONSTANT' },
    LEARN_RATE = float64_value,
    EARLY_STOP = { TRUE },
    MIN_REL_PROGRESS = float64_value,
    DATA_SPLIT_METHOD = { 'AUTO_SPLIT'},
    DATA_SPLIT_EVAL_FRACTION = float64_value,
    DATA_SPLIT_COL = string_value,
    LS_INIT_LEARN_RATE = float64_value,
    WARM_START = { FALSE },
    AUTO_CLASS_WEIGHTS = { TRUE  },
    CLASS_WEIGHTS = struct_array,
    ENABLE_GLOBAL_EXPLAIN = { FALSE },
    CALCULATE_P_VALUES = { FALSE },
    FIT_INTERCEPT = { FALSE },
    CATEGORY_ENCODING_METHOD = { 'ONE_HOT_ENCODING`, 'DUMMY_ENCODING'
})];
```

The key options that can be specified in the CREATE MODEL statement are as follows:

- MODEL_TYPE: Specifies the required model type (for example, linear or logistic regression).

- INPUT_LABEL_COLS: Defines the label column names in the training data.

- OPTIMIZE_STRATEGY: Selects the approach for training linear regression models:

  - AUTO_STRATEGY: Chooses the training approach based on several conditions:

    - The batch_gradient_descent strategy is employed if either l1_reg or warm_start is specified

    - batch_gradient_descent is also used if the overall cardinality of training features surpasses 10,000

    - When overfitting may be an issue, specifically when the number of training samples is less than 10 times the total cardinality, batch_gradient_descent is chosen

    - For all other scenarios, the NORMAL_EQUATION strategy is implemented

  - BATCH_GRADIENT_DESCENT: Engages the batch gradient descent method for model training, optimizing the loss function through the use of the gradient function.

  - NORMAL_EQUATION: Derives the least square solution for the linear regression issue using an analytical formula. The use of the NORMAL_EQUATION strategy is not permissible in the following situations:

    - l1_reg is defined

    - warm_start is defined

    - The total cardinality of training features exceeds 10,000

- L1_REG: Sets the amount of L1 regularization applied.

- L2_REG: Sets the amount of L2 regularization applied.

- MAX_ITERATIONS: Determines the maximum number of training iterations or steps.

- LEARN_RATE_STRATEGY: Selects the strategy for specifying the learning rate during training.

- LEARN_RATE: Defines the learning rate for gradient descent.

- EARLY_STOP: Indicates whether training should stop after the first iteration with minimal relative loss improvement.

- MIN_REL_PROGRESS: Sets the minimum relative loss improvement to continue training.

- DATA_SPLIT_METHOD: Chooses the method for splitting input data into training and evaluation sets. The options here are 'AUTO_SPLIT', 'RANDOM', 'CUSTOM', 'SEQ', and 'NO_SPLIT'.

- DATA_SPLIT_EVAL_FRACTION: Specifies the fraction of data used for evaluation with 'RANDOM' and 'SEQ' splits.

- DATA_SPLIT_COL: Identifies the column used to split the data.

- LS_INIT_LEARN_RATE: Sets the initial learning rate for the 'LINE_SEARCH' strategy.

- WARM_START: Retrains a model with new training data, new model options, or both.

- AUTO_CLASS_WEIGHTS: Balances class labels using weights for each class in inverse proportion to the frequency of that class.

- CLASS_WEIGHTS: Defines the weights to use for each class label.

- ENABLE_GLOBAL_EXPLAIN: Computes global explanations using Explainable AI for global feature importance evaluation.

- CALCULATE_P_VALUES: Computes p-values and standard errors during training.

- FIT_INTERCEPT: Fits an intercept to the model during training.

- CATEGORY_ENCODING_METHOD: Specifies the encoding method to use on non-numeric features.

### Creating deep neural network models and wide-and-deep models

Here's the syntax for creating deep learning models, along with the different required and optional arguments you need to provide as part of the query:

```
{CREATE OR REPLACE MODEL} model_name
[OPTIONS(MODEL_TYPE= {'DNN_CLASSIFIER'},
        ACTIVATION_FN = { 'RELU' },
        AUTO_CLASS_WEIGHTS = { TRUE | FALSE },
        BATCH_SIZE = int64_value,
        CLASS_WEIGHTS = struct_array,
        DROPOUT = float64_value,
        EARLY_STOP = { TRUE | FALSE },
        HIDDEN_UNITS = int_array,
        L1_REG = float64_value,
        L2_REG = float64_value,
        LEARN_RATE = float64_value,
        INPUT_LABEL_COLS = string_array,
        MAX_ITERATIONS = int64_value,
        MIN_REL_PROGRESS = float64_value,
        OPTIMIZER={'ADAGRAD'},
        WARM_START = { FALSE },
        DATA_SPLIT_METHOD={'AUTO_SPLIT'},
        DATA_SPLIT_EVAL_FRACTION = float64_value,
        DATA_SPLIT_COL = string_value,
        ENABLE_GLOBAL_EXPLAIN = { FALSE },
        INTEGRATED_GRADIENTS_NUM_STEPS = int64_value,
        TF_VERSION = { '2.8.0' })];
```

The following options can be specified as part of the model creation request:

- `model_name`: The name of the BQML model you're creating or replacing.

- `model_type`: Specifies the type of model, either `'DNN_CLASSIFIER'` or `'DNN_REGRESSOR'`.

- `activation_fn`: For DNN model types, this specifies the activation function of the neural network. The options are `'RELU'`, `'RELU6'`, `'CRELU'`, `'ELU'`, `'SELU'`, `'SIGMOID'`, and `'TANH'`.

- `auto_class_weights`: Specifies whether to balance class labels using weights for each class in inverse proportion to the frequency of that class. Use only with the `DNN_CLASSIFIER` model.

- `batch_size`: For DNN model types, this specifies the mini-batch size of samples that are fed to the neural network.

- `class_weights`: The weights to use for each class label. This option cannot be specified if `AUTO_CLASS_WEIGHTS` is TRUE.

- `data_split_method`: The method to split input data into training and evaluation sets. The options are `'AUTO_SPLIT'`, `'RANDOM'`, `'CUSTOM'`, `'SEQ'`, and `'NO_SPLIT'`.

- `data_split_eval_fraction`: Used with `'RANDOM'` and `'SEQ'` splits. It specifies the fraction of the data used for evaluation.

- `data_split_col`: Identifies the column used to split the data.

- `dropout`: For DNN model types, this specifies the dropout rate of units in the neural network.

- `early_stop`: Whether training should stop after the first iteration in which the relative loss improvement is less than the value specified for `MIN_REL_PROGRESS`.

- `enable_global_explain`: Specifies whether to compute global explanations using Explainable AI to evaluate global feature importance to the model.

- `hidden_units`: For DNN model types, this specifies the hidden layers of the neural network.

- `input_label_cols`: The label column name(s) in the training data.

- `integrated_gradients_num_steps`: Specifies the number of steps to sample between the example being explained and its baseline for approximating the integral in integrated gradients attribution methods.

- `l1_reg`: The L1 regularization strength of the optimizer.

- `l2_reg`: The L2 regularization strength of the optimizer.

- `learn_rate`: The initial learning rate for training.

- `max_iterations`: The maximum number of training iterations.

- `optimizer`: For DNN model types, this specifies the optimizer for training the model. The options are `'ADAGRAD'`, `'ADAM'`, `'FTRL'`, `'RMSPROP'`, and `'SGD'`.

- `warm_start`: Whether to retrain a model with new training data, new model options, or both.

- `tf_version`: Specifies the TensorFlow version for model training.

### Creating boosted tree and random forest models

Here's the syntax for creating boosted tree and random forest models, along with different required and optional arguments you need to provide as part of the query:

```
{CREATE OR REPLACE MODEL} model_name
[OPTIONS(MODEL_TYPE = { 'BOOSTED_TREE_CLASSIFIER' },
        BOOSTER_TYPE = {'GBTREE' },
        NUM_PARALLEL_TREE = int64_value,
        DART_NORMALIZE_TYPE = {'TREE' },
        TREE_METHOD={'AUTO' },
        MIN_TREE_CHILD_WEIGHT = int64_value,
        COLSAMPLE_BYTREE = float64_value,
        COLSAMPLE_BYLEVEL = float64_value,
        COLSAMPLE_BYNODE = float64_value,
        MIN_SPLIT_LOSS = float64_value,
        MAX_TREE_DEPTH = int64_value,
        SUBSAMPLE = float64_value,
        AUTO_CLASS_WEIGHTS = { TRUE },
        CLASS_WEIGHTS = struct_array,
        INSTANCE_WEIGHT_COL = string_value,
        L1_REG = float64_value,
        L2_REG = float64_value,
        EARLY_STOP = { TRUE },
        LEARN_RATE = float64_value,
        INPUT_LABEL_COLS = string_array,
        MAX_ITERATIONS = int64_value,
        MIN_REL_PROGRESS = float64_value,
        DATA_SPLIT_METHOD = {'AUTO_SPLIT'},
        DATA_SPLIT_EVAL_FRACTION = float64_value,
        DATA_SPLIT_COL = string_value,
        ENABLE_GLOBAL_EXPLAIN = { TRUE},
        XGBOOST_VERSION = {'1.1'})];
```

Here are the options that can be specified as part of the model creation request:

- MODEL_TYPE: Specifies whether the model is a boosted tree classifier, boosted tree regressor, random forest classifier, or random forest regressor. The options are 'BOOSTED_TREE_CLASSIFIER', 'BOOSTED_TREE_REGRESSOR', 'RANDOM_FOREST_CLASSIFIER', and 'RANDOM_FOREST_REGRESSOR'.

- BOOSTER_TYPE (applicable only for Boosted_Tree_Models): Specifies the type of booster used for the boosted tree model. **GBTREE** stands for **Gradient Boosting Tree** and **DART** stands for **Dropouts meet Multiple Additive Regression Trees**.

- NUM_PARALLEL_TREE: Specifies the number of parallel trees to grow. Larger numbers can lead to improved performance but can also increase training time and memory usage.

- DART_NORMALIZE_TYPE (applicable only for Boosted_Tree_Models): Specifies the normalization method used for the **DART** booster. 'TREE' means normalization by the number of dropped trees in the boosting process and 'FOREST' means normalization by the total number of trees in the forest.

- TREE_METHOD: Specifies the method used to construct each decision tree in the ensemble. 'AUTO' means that the algorithm will choose the best method based on the data, 'EXACT' means exact greedy algorithm, 'APPROX' means approximate greedy algorithm, and 'HIST' means histogram-based algorithm.

- MIN_TREE_CHILD_WEIGHT: Specifies the minimum sum of instance weights required in a child node of a tree. If the sum is below this value, the node will not be split.

- COLSAMPLE_BYTREE: Specifies the fraction of columns to be randomly sampled for each tree.

- COLSAMPLE_BYLEVEL: Specifies the fraction of columns to be randomly sampled for each level of a tree.

- COLSAMPLE_BYNODE: Specifies the fraction of columns to be randomly sampled for each split node of a tree.

- MIN_SPLIT_LOSS: Specifies the minimum loss reduction required to split a node.

- MAX_TREE_DEPTH: Specifies the maximum depth of each tree.

- SUBSAMPLE: Specifies the fraction of training instances to be randomly sampled for each tree.

- AUTO_CLASS_WEIGHTS: If set to TRUE, the algorithm will automatically determine the weights to be assigned to each class based on the data.

- CLASS_WEIGHTS: Specifies the weight to be assigned to each class. This can be used to balance the data if the classes are imbalanced.

- INSTANCE_WEIGHT_COL: Specifies the name of the column containing the instance weights.

- L1_REG: Specifies the L1 regularization parameter.

- `L2_REG`: Specifies the L2 regularization parameter.

- `EARLY_STOP`: If set to `TRUE`, the training process will stop early if the performance improvement falls below a certain threshold. The options are `TRUE` and `FALSE`.

- `LEARN_RATE` (applicable only for `Boosted_Tree_Models`): Specifies the learning rate, which controls the step size at each iteration of the boosting process.

- `INPUT_LABEL_COLS`: Specifies the names of the columns containing the input features and the label.

- `MAX_ITERATIONS` (applicable only for `Boosted_Tree_Models`): Specifies the maximum number of boosting iterations to perform.

- `MIN_REL_PROGRESS`: Specifies the minimum relative progress required to continue the training process.

- `DATA_SPLIT_METHOD`: Specifies the method used to split the data into training and validation sets. `'AUTO_SPLIT'` means that the algorithm will automatically split the data, `'RANDOM'` means random splitting, `'CUSTOM'` means user-defined splitting, `'SEQ'` means sequential splitting, and `'NO_SPLIT'` means no splitting (use all data for training).

- `DATA_SPLIT_EVAL_FRACTION`: Specifies the fraction of the data to be used for validation when splitting the data.

- `DATA_SPLIT_COL`: Specifies the name of the column used to split the data.

- `ENABLE_GLOBAL_EXPLAIN`: If set to `TRUE`, the algorithm will compute global feature importance scores. The options are `TRUE` and `FALSE`.

- `XGBOOST_VERSION`: Specifies the version of XGBoost to be used.

## Importing models

BQML also allows you to import deep learning models trained outside BigQuery. This is an extremely useful feature because it gives you the flexibility to train models using a more custom setup outside BigQuery and yet be able to use BigQuery's compute infrastructure for inference.

Here is how you can use the import feature:

```
{CREATE OR REPLACE MODEL} model_name
[OPTIONS(MODEL_TYPE = {'TENSORFLOW'} ,
MODEL_PATH = string_value)]
```

Here are the available options as part of the import feature:

- `MODEL_TYPE`: Specifies whether the model is TensorFlow, TensorFlow Lite, or an ONNX model. The options are `'TENSORFLOW'`, `'ONNX'`, and `'TENSORFLOW_LITE'`.

- `MODEL_PATH`: Provides the Cloud Storage URI of the model to import into BQML.

## k-means models

Here's the syntax for creating k-means models, along with different required and optional arguments you need to provide as part of the query:

```
{CREATE OR REPLACE MODEL} model_name
[OPTIONS(MODEL_TYPE = { 'KMEANS' },
    NUM_CLUSTERS = int64_value,
    KMEANS_INIT_METHOD = { 'RANDOM' },
    KMEANS_INIT_COL = string_value,
    DISTANCE_TYPE = { 'EUCLIDEAN' | 'COSINE' },
    STANDARDIZE_FEATURES = { TRUE },
    MAX_ITERATIONS = int64_value,
    EARLY_STOP = { TRUE },
    MIN_REL_PROGRESS = float64_value,
    WARM_START = { FALSE })];
```

Let's look at the options that can be specified as part of the model creation query:

- MODEL_TYPE: Specifies the type of model. This option is required.

- NUM_CLUSTERS (optional): For a k-means model, this specifies the number of clusters to identify in the input data. The default value is log10(n), where n is the number of training examples.

- KMEANS_INIT_METHOD (optional): For a k-means model, this specifies the method of initializing the clusters. The default value is 'RANDOM'. The options are 'RANDOM', 'KMEANS++', and 'CUSTOM'.

- KMEANS_INIT_COL (optional): For a k-means model, this identifies the column that will be used to initialize the centroids. This option can only be specified when KMEANS_INIT_METHOD has a value of CUSTOM. The corresponding column must be of the BOOL type, and the NUM_CLUSTERS model option must be present in the query and its value must equal the total number of TRUE rows in this column. BQML cannot use this column as a feature and excludes it from features automatically.

- DISTANCE_TYPE (optional): For a k-means model, this specifies the type of metric to compute the distance between two points. The default value is 'EUCLIDEAN'.

- STANDARDIZE_FEATURES (optional): For a k-means model, this specifies whether to standardize numerical features. The default value is TRUE.

- MAX_ITERATIONS (optional): The maximum number of training iterations or steps. The default value is 20.

- EARLY_STOP (optional): Whether training should stop after the first iteration in which the relative loss improvement is less than the value specified for MIN_REL_PROGRESS. The default value is TRUE.

- MIN_REL_PROGRESS (optional): The minimum relative loss improvement that is necessary to continue training when EARLY_STOP is set to TRUE. For example, a value of 0.01 specifies that each iteration must reduce the loss by 1% for training to continue. The default value is 0.01.

- WARM_START (optional): Whether to retrain a model with new training data, new model options, or both. Unless explicitly overridden, the initial options used to train the model are used for the warm start run. The value of MODEL_TYPE and the training data schema must remain constant in a warm start model's retraining. The default value is FALSE.

Now let's look at the support BQML offers for hyperparameter tuning.

## Hyperparameter tuning with BQML

BQML allows you to fine-tune hyperparameters when building ML models through the use of CREATE MODEL statements. This process, known as hyperparameter tuning, is a commonly employed method for enhancing model accuracy by finding the ideal set of hyperparameters.

Here's an example BigQuery SQL statement:

```
{CREATE OR REPLACE MODEL} model_name
OPTIONS(Existing Training Options,
    NUM_TRIALS = int64_value, [, MAX_PARALLEL_TRIALS = int64_value ]
    [, HPARAM_TUNING_ALGORITHM = { 'VIZIER_DEFAULT' | 'RANDOM_SEARCH' |
'GRID_SEARCH' } ]
    [, hyperparameter={HPARAM_RANGE(min, max) | HPARAM_
CANDIDATES([candidates]) }... ]
    [, HPARAM_TUNING_OBJECTIVES = { 'R2_SCORE' | 'ROC_AUC' | ... } ]
    [, DATA_SPLIT_METHOD = { 'AUTO_SPLIT' | 'RANDOM' | 'CUSTOM' | 'SEQ'
| 'NO_SPLIT' } ]
    [, DATA_SPLIT_COL = string_value ]
    [, DATA_SPLIT_EVAL_FRACTION = float64_value ]
    [, DATA_SPLIT_TEST_FRACTION = float64_value ]
) AS query_statement
```

Let's look at the options that can be specified as part of the model creation query:

- NUM_TRIALS

  - Description: This determines the maximum number of submodels to train. Tuning will cease after training num_trials submodels or upon search space exhaustion. The maximum value is 100.

  - Arguments: int64_value must be an INT64 value ranging from 1 to 100.

> **Note**
>
> It is suggested to use at least (`num_hyperparameters` * 10) trials for model tuning.3

- `MAX_PARALLEL_TRIALS`

  - Description: This represents the maximum number of trials to run concurrently. The default value is 1, while the maximum value is 5.

  - Arguments: `int64_value` must be an `INT64` value ranging from 1 to 5.

> **Note**
>
> A larger `max_parallel_trials` value can speed up hyperparameter tuning, but it may compromise the final model's quality for the `VIZIER_DEFAULT` tuning algorithm as parallel trials cannot benefit from concurrent training results.

- `HPARAM_TUNING_ALGORITHM`

  - Description: This determines the algorithm for hyperparameter tuning and supports the following values:

    - `VIZIER_DEFAULT` (default and recommended): Uses the default Vertex AI Vizier algorithm, which combines advanced search algorithms such as Bayesian optimization with Gaussian processes and employs transfer learning to utilize previously tuned models.

    - `RANDOM_SEARCH`: Employs random search to explore the search space.

    - `GRID_SEARCH`: Utilizes grid search to explore the search space. This is only available when every hyperparameter's search space is discrete.

- `HYPERPARAMETER`

  Syntax: `hyperparameter={HPARAM_RANGE(min, max) | HPARAM_CANDIDATES([candidates]) }...`

  This parameter configures a hyperparameter's search space. Refer to the hyperparameters and objectives for each model type to find out which tunable hyperparameters are supported:

  - `HPARAM_RANGE(min, max)`: Specifies the continuous search space for a hyperparameter – for example, `learn_rate = HPARAM_RANGE(0.0001, 1.0)`

  - `HPARAM_CANDIDATES([candidates])`: Specifies a hyperparameter with discrete values – for example, `OPTIMIZER=HPARAM_CANDIDATES(['adagrad', 'sgd', 'ftrl'])`

- `HPARAM_TUNING_OBJECTIVES`

This parameter specifies objective metrics for the model. The candidates are a subset of model evaluation metrics. Only one objective is supported currently. Refer to *Table 6.7*, which shows each model type, to see the supported hyperparameters and tuning objectives.

| Model Type | Hyperparameter Objectives | Hyperparameter |
|---|---|---|
| LINEAR_REG | • mean_absolute_error<br>• mean_squared_error<br>• mean_squared_log_error<br>• median_absolute_error<br>• r2_score (default)<br>• explained_variance | • l1_reg<br>• l2_reg |
| LOGISTIC_REG | • precision<br>• recall<br>• accuracy<br>• f1_score<br>• log_loss<br>• roc_auc (default) | • l1_reg<br>• l2_reg |
| KMEANS | • davies_bouldin_index | num_clusters |
| MATRIX_<br>FACTORIZATION<br>(implicit/<br>explicit) | • mean_average_precision (explicit model)<br>• mean_squared_error (implicit/explicit)<br>• normalized_discounted_cumulative_gain (explicit model)<br>• average_rank (explicit model) | • num_factors<br>• l2_reg<br>• wals_alpha(implicit model only) |

| Model Type | Hyperparameter Objectives | Hyperparameter |
|---|---|---|
| DNN_CLASSIFIER | • precision<br>• recall<br>• accuracy<br>• f1_score<br>• log_loss<br>• roc_auc (default) | • batch_size<br>• dropout<br>• hidden_units<br>• learn_rate<br>• optimizer<br>• l1_reg |
| DNN_REGRESSOR | • mean_absolute_error<br>• mean_squared_error<br>• mean_squared_log_error<br>• median_absolute_error<br>• r2_score (default)<br>• explained_variance | • l2_reg<br>• activation_fn |
| BOOSTED_TREE_<br>CLASSIFIER | • precision<br>• recall<br>• accuracy<br>• f1_score<br>• log_loss<br>• roc_auc (default) | • learn_rate<br>• l1_reg<br>• l2_reg<br>• dropout<br>• max_tree_depth<br>• subsample |
| BOOSTED_TREE_<br>REGRESSOR | • mean_absolute_error<br>• mean_squared_error<br>• mean_squared_log_error<br>• median_absolute_error<br>• r2_score (default)<br>• explained_variance | • min_split_loss<br>• num_parallel_tree<br>• min_tree_child_<br>weight<br>• colsample_bytree<br>• colsample_bylevel<br>• colsample_bynode<br>• booster_type<br>• dart_normalize_type<br>• tree_method |

| Model Type | Hyperparameter Objectives | Hyperparameter |
|---|---|---|
| RANDOM_FOREST_CLASSIFIER | • precision<br>• recall<br>• accuracy<br>• f1_score<br>• log_loss<br>• roc_auc (default) | • l1_reg<br>• l2_reg<br>• max_tree_depth<br>• subsample<br>• min_split_loss<br>• num_parallel_tree |
| RANDOM_FOREST_REGRESSOR | • mean_absolute_error<br>• mean_squared_error<br>• mean_squared_log_error<br>• median_absolute_error<br>• r2_score (default)<br>• explained_variance | • min_tree_child_weight<br>• colsample_bytree<br>• colsample_bylevel<br>• colsample_bynode<br>• tree_method |

Table 6.7 – Supported hyperparameter objectives by model type

```
https://cloud.google.com/bigquery/docs/reference/standard-sql/
bigqueryml-hyperparameter-tuning
```

Now let's look at BQML features that you can use when trying to evaluate ML models.

# Evaluating trained models

Once the BQML model has been trained, you will want to evaluate the key performance statistics, depending on the type of model. You can do so by using the ML.EVALUATE function, as shown here:

```
ML.EVALUATE(MODEL model_name
          [, {TABLE table_name | (query_statement)}]
          [, STRUCT<threshold FLOAT64,
                    perform_aggregation BOOL,
                    horizon INT64,
                    confidence_level FLOAT64> settings])])])
```

Let's look at the options you can specify as part of the evaluation query:

- `model_name`: The name of the model being evaluated
- `table_name` (optional): The name of the table containing the evaluation data
- `query_statement` (optional): The query used to generate the evaluation data
- `threshold` (optional): A custom threshold value for binary-class classification models that's used during evaluation
- `perform_aggregation` (optional): A Boolean value that identifies the level of evaluation for forecasting accuracy
- `horizon` (optional): The number of forecasted time points against which evaluation metrics are computed
- `confidence_level` (optional): The percentage of future values that fall within the prediction interval

The output of the `ML.Evaluate` function depends on the type of model being evaluated:

| Model Type | Returned Fields |
| --- | --- |
| Regression models | <ul><li>`mean_absolute_error`</li><li>`mean_squared_error`</li><li>`mean_squared_log_error`</li><li>`median_absolute_error`</li><li>`r2_score`</li><li>`explained_variance`</li></ul> |
| Classification models | <ul><li>`precision`</li><li>`recall`</li><li>`accuracy`</li><li>`f1_score`</li><li>`log_loss`</li><li>`roc_auc`</li></ul> |
| k-means model | <ul><li>`Davies-Bouldin index`</li><li>`Mean squared distance`</li></ul> |

| Model Type | Returned Fields |
|---|---|
| Matrix factorization model with implicit feedback | • `mean_average_precision`<br>• `mean_squared_error`<br>• `normalized_discounted_cumulative_gain`<br>• `average_rank` |
| Matrix factorization model with explicit feedback | • `mean_absolute_error`<br>• `mean_squared_error`<br>• `mean_squared_log_error`<br>• `median_absolute_error`<br>• `r2_score` |
| PCA model | • `explained_variance`<br>• `total_explained_variance_ratio` |
| Time series ARIMA_PLUS or ARIMA_PLUS_XREG model with input data and perform_aggregation = false | • `time_series_id_col` or `time_series_id_cols`<br>• `time_series_timestamp_col`<br>• `time_series_data_col`<br>• `forecasted_time_series_data_col`<br>• `lower_bound`<br>• `upper_bound`<br>• `absolute_error`<br>• `absolute_percentage_error` |
| Time series ARIMA_PLUS or ARIMA_PLUS_XREG model with input data and perform_aggregation = true | • `time_series_id_col` or `time_series_id_cols`<br>• `mean_absolute_error`<br>• `mean_squared_error`<br>• `root_mean_squared_error`<br>• `mean_absolute_percentage_error`<br>• `symmetric_mean_absolute_percentage_error` |

| Model Type | Returned Fields |
|---|---|
| | • `time_series_id_col` or `time_series_id_cols` |
| | • `non_seasonal_p` |
| | • `non_seasonal_d` |
| | • `non_seasonal_q` |
| | • `has_drift` |
| Time series `ARIMA_PLUS` model without input data | • `log_likelihood` |
| | • `AIC` |
| | • `variance` |
| | • `seasonal_periods` |
| | • `has_holiday_effect` |
| | • `has_spikes_and_dips` |
| | • `has_step_change` |
| | • `mean_absolute_error` |
| Autoencoder model | • `mean_squared_error` |
| | • `mean_squared_log_error` |
| Remote model | • `remote_eval_metrics` |

Table 6.8 – ML.Evaluate Output

In the next section we look at how you can use your BQML models for inference.

## Doing inference with BQML

In supervised ML, the ultimate goal is to use a trained model to make predictions on new data. BQML provides the `ML.PREDICT` function for this purpose. Using this function, you can easily predict outcomes by supplying new data to a trained model. The `ML.PREDICT` function can be used during model creation, after model creation, or after a failure, so long as at least one iteration has been completed. The function returns a table with the same number of rows as the input table, and it includes all columns from the input table and all output columns from the model, with the output column names prefixed with `predicted_`.

```
ML.PREDICT(MODEL model_name,
          {TABLE table_name | (query_statement)}
          [, STRUCT<threshold FLOAT64,
          keep_original_columns BOOL> settings)])
```

The output fields that are included in the response of the `ML.PREDICT` function depends on the type of model being used:

| Model Type | Output Columns |
|---|---|
| Linear regression<br><br>Boosted tree regressor<br><br>Random forest regressor<br><br>DNN regressor | `predicted_<label_column_name>` |
| Binary logistic regression<br><br>Boosted tree classifier<br><br>Random forest classifier<br><br>DNN classifier<br><br>Multiclass logistic regression | `predicted_<label_column_name>,`<br>`predicted_<label_column_name>_probs` |
| k-means | `centroid_id, nearest_centroids_distance` |
| PCA | `principal_component_<index>,`<br><br>input columns (if keep_original_columns is set to true) |
| Autoencoder | `latent_col_<index>,`<br><br>input columns |
| TensorFlow Lite | The output of the TensorFlow Lite model's predict method |
| Remote models | The output columns containing all Vertex AI endpoint output fields, and a remote_model_status field containing status messages from the Vertex AI endpoint |
| ONNX models | The output of the ONNX model's predict method |
| XGBoost models | The output of the XGBoost model's predict method |

Table 6.9 – ML.Predict Output

Now let's work through a hands-on exercise where we use BQML to train an ML model and use it for generating predictions.

# User exercise

Refer to the notebook in *Chapter 6, Low-Code Options for Building ML Models*, in this book's GitHub repository for a hands-on exercise around training a BQML model. In this exercise, you will use one of the public datasets available in BigQuery to train a model to predict the likelihood of a customer defaulting on their loan next month.

# Summary

BQML is a powerful tool for data scientists and analysts who want to train ML models with ease while using a low-code option to build and deploy models in GCP. With BQML, users can leverage the power of BigQuery to quickly and easily create models without needing to write complex code.

In this chapter, we explored the features and benefits of BQML. We saw how it provides a simple and intuitive interface for training models using SQL queries. We also explored some of the key features of BQML, including the ability to perform data preprocessing and feature engineering directly within BigQuery, as well as the ability to evaluate model performance through native evaluation functions.

One of the key advantages of BQML is its integration with BigQuery, which makes it easy to scale and manage large datasets. This makes it a great option for companies and organizations that are dealing with massive amounts of data and need to quickly build and deploy models.

Another advantage of BQML is its support for a wide range of ML models, including linear regression, logistic regression, k-means clustering, and more. This makes it a versatile tool that can be used for a variety of use cases, from predicting customer churn to clustering data for analysis.

We also discussed some of the limitations of BQML. For example, while it provides a low-code option for building and deploying models, it may not be suitable for more complex use cases that require custom models or extensive feature engineering. Additionally, while BQML provides a range of metrics for evaluating model performance, users may need to do additional analysis to fully understand the effectiveness of their models.

Despite these limitations, BQML is a powerful tool for data scientists and analysts who want to quickly and easily build and deploy ML models. Its integration with BigQuery and other GCP services makes it a great option for companies and organizations that need to work with large amounts of data, while its support for a wide range of models and metrics makes it a versatile tool for a variety of use cases.

Overall, BQML is a valuable addition to the suite of ML tools available in GCP. Its low-code interface, integration with BigQuery, and support for a wide range of models make it a great option for data scientists and analysts who want to focus on their data and insights, rather than complex code and infrastructure. With BQML, users can quickly and easily build and deploy models, enabling them to extract valuable insights from their data and make data-driven decisions.

In the next chapter, we will look at how to train fully custom TensorFlow deep learning models on Vertex AI using its serverless training features. This chapter will also do a deep dive into building the model using TensorFlow, packaging it for submission to Vertex AI, monitoring the training progress, and evaluating the trained model.

# Training Fully Custom ML Models with Vertex AI

In the previous chapters, we learned about training no-code (Auto-ML) as well as low-code (BQML) **Machine Learning** (ML) models with minimum technical expertise required. These solutions are really handy when it comes to solving common ML problems. However, sometimes the problem or data itself is so complex that it requires the development of custom **Artificial Intelligence** (AI) models, in most cases large deep learning-based models. Working on custom models requires a significant level of technical expertise in the fields of ML, deep learning, and AI. Sometimes, even with this expertise, it becomes really difficult to manage training and experiments of large-scale custom deep learning models due to a lack of resources, compute, and proper metadata tracking mechanisms.

To make the lives of ML developers easier, Vertex AI provides a managed environment for launching large-scale custom training jobs. Vertex AI-managed jobs let us track useful metadata, monitor jobs through the Google Cloud console UI, and launch large-scale batch inference jobs without the need to actively monitor them. In this chapter, we will learn how to work with custom deep learning-based models on Google Vertex AI. Specifically, we will cover the following topics:

- Building a basic deep learning model with TensorFlow
- Packaging a model to submit to Vertex AI as a training job
- Monitoring model training progress
- Evaluating trained models

## Technical requirements

This chapter requires basic-level knowledge of the deep learning framework TensorFlow and neural networks. Code artifacts can be found in the following GitHub repo – https://github.com/PacktPublishing/The-Definitive-Guide-to-Google-Vertex-AI/tree/main/Chapter07

# Building a basic deep learning model with TensorFlow

**TensorFlow**, or **TF** for short, is an end-to-end platform for building ML models. The main focus of the TensorFlow framework is to simplify the development, training, evaluation, and deployment of deep neural networks. When it comes to working with unstructured data (such as images, videos, audio, etc.), neural network-based solutions have achieved significantly better results than traditional ML approaches that mostly rely on handcrafted features. Deep neural networks are good at understanding complex patterns from high-dimensional data points (for example, an image with millions of pixels). In this section, we will develop a basic neural network-based model using TensorFlow. In the next few sections, we will see how Vertex AI can help with setting up scalable and systemic training/tuning of such custom models.

> **Important Note**
>
> It is important to note that TensorFlow is not the only ML framework that Vertex AI supports. Vertex AI supports many different ML frameworks and open-source projects including Pytorch, Spark and XGBoost. Pytorch is one of the fastest growing ML frameworks and with Vertex AI's Pytorch integrations, we can easily train, deploy and orchestrate PyTorch models in production. Vertex AI provides prebuilt training and serving containers and also supports optimized distributed training of PyTorch models. Similarly, Vertex AI provides prebuilt training, serving and explainability features for multiple ML frameworks including XGBoost, TensorFlow, Pytorch and Scikit-learn.

## Experiment – converting black-and-white images into color images

In this experiment, we will develop a TensorFlow-based deep learning model that takes black-and-white images as input and converts them into color images. As this exercise, requires developing a custom model, we will start our initial development work on a Jupyter Notebook. The first step is to create a user-managed Jupyter Notebook inside Vertex AI Workbench using a preconfigured TensorFlow image. More details on how to successfully create a Vertex AI Workbench notebook instance can be found in *Chapter 4*, *Vertex AI Workbench*. Next, let's launch one Jupyter Notebook from the JupyterLab application. We are now all set to start working on our experiment.

We will start with importing useful libraries (prebuilt Python packages) in the first cell of our notebook. In this experiment, we will be using the following Python libraries – `numpy` for multi-dimensional array manipulation, TensorFlow for developing a deep learning model, OpenCV (or `cv2`) for image manipulation, and `matplotlib` for plotting images or graphs:

```
import numpy as np
import tensorflow
import glob
import cv2
```

```
import matplotlib.pyplot as plt
%matplotlib inline
```

We will need an image dataset with at least a few thousand images to train and test our model. In this experiment, we will work with the **Oxford-IIIT Pet** dataset, which is a public and free-to-use dataset. This dataset consists of around 7k pet images from more than 30 different annotated categories. The dataset can be downloaded from the following website: `https://www.robots.ox.ac.uk/~vgg/data/pets/`.

We can also download this dataset using the following commands in our terminal:

```
wget https://thor.robots.ox.ac.uk/~vgg/data/pets/images.tar.gz
wget https://thor.robots.ox.ac.uk/~vgg/data/pets/annotations.tar.gz
```

Once the downloads are complete, put the zipped files in the same directory as our notebook. Now, let's create a **data** folder, extract all images into it, and extract the annotations into another folder. We can do this in a notebook cell using the following command (the ! sign in a notebook cell lets us run terminal commands from within Jupyter Notebook):

```
!mkdir data
!tar -xf images.tar.gz -C data
!mkdir labels
!tar -xf annotations.tar.gz -C labels
```

As our current experiment is concerned with converting black-and-white images into colored versions, we will not be using annotations. Now, let's quickly verify in a new cell whether we have all the images successfully copied into the **data** folder:

```
all_image_files = glob.glob("data/images/*.jpg")
print("Total number of image files : ", \
    len(all_image_files))
```

Here, the `glob` module helps us by listing all the `.jpg` image paths inside the data directory. The length of this list will be equal to the number of images. The preceding code should print the following output:

```
Total number of image files :  7390
```

Now that we have successfully downloaded and extracted data, let's check a few images to be sure everything is fine. The following code block will plot a few random images with their annotations:

```
for i in range(3):
    plt.figure(figsize=(13, 13))
    for j in range(6):
        img_path = np.random.choice(all_image_files)
        img = cv2.imread(img_path)
        img_class = img_path.split("/")[-1].split("_")[0]
```

```
        img = cv2.cvtColor(img, cv2.COLOR_BGR2RGB)
        plt.subplot(660 + 1 + j)
        plt.imshow(img)
        plt.axis('off')
        plt.title(img_class)
    plt.show()
```

Here, we are extracting the image class (or annotation) from the image path name itself as all images have the pet category in their filenames. The output of the preceding code should look something like shown in *Figure 7.1*.

Figure 7.1 – A few samples from the pet dataset

Now that we have verified our dataset, let's split these images into three sets – train, validation, and test – as we usually do for training/validating and testing ML models. We will keep 60% of the images for training, 20% for validation, and the remaining 20% for testing. One simple way to do these splits is as follows:

```
train_files = all_image_files[:int( \
    len(all_image_files)*0.6)]
validation_files = all_image_files[int( \
    len(all_image_files)*0.6):int(len(all_image_files)*0.8)]
test_files = all_image_files[int( \
    len(all_image_files)*0.8):]
print(len(train_files), len(validation_files), \
    len(test_files))
```

The main focus of this experiment is to develop a deep learning model that converts black-and-white images into color images. To learn this mapping, the model will require pairs of black-and-white and corresponding color versions in order to learn this mapping. Our dataset already has color images. We will utilize the OpenCV library to convert them into grayscale (black-and-white) images and use them as input in our model. We will compare the output with their color versions. Another important thing to keep in mind is that our deep learning model will take fixed-size images as inputs, so we also need to bring all input images into a common resolution. In our experiment, we will change all of our images to have an 80x80 resolution. We already have training, validation, and test splits of the image paths. We can now read those files and prepare data for training, validation, and testing purposes. The following code blocks can be used to prepare the dataset as described previously.

Let's first define empty lists for storing training, validation, and test data, respectively:

```
train_x = []
train_y = []
val_x = []
val_y = []
test_x = []
test_y = []
```

Here, we read training images, resize them to the required size, create a black-and-white version of each of them, and store them as target images:

```
for file in train_files:
    try:
        img = cv2.imread(file)
        img = cv2.resize(img, (80,80))
        color_img = cv2.cvtColor(img, cv2.COLOR_BGR2RGB)
        black_n_white_img = cv2.cvtColor(color_img, \
            cv2.COLOR_RGB2GRAY)
    except:
        continue
    train_x.append((black_n_white_img-127.5)/127.5)
    train_y.append((color_img-127.5)/127.5)
```

Similarly, we repeat the same process for validation files:

```
for file in validation_files:
    try:
        img = cv2.imread(file)
        img = cv2.resize(img, (80,80))
        color_img = cv2.cvtColor(img, cv2.COLOR_BGR2RGB)
        black_n_white_img = cv2.cvtColor(color_img, \
            cv2.COLOR_RGB2GRAY)
    except:
```

```
        continue
    val_x.append((black_n_white_img-127.5)/127.5)
    val_y.append((color_img-127.5)/127.5)
```

Now, prepare test files as well in a similar way:

```
for file in test_files:
    try:
        img = cv2.imread(file)
        img = cv2.resize(img, (80,80))
        color_img = cv2.cvtColor(img, cv2.COLOR_BGR2RGB)
        black_n_white_img = cv2.cvtColor(color_img, \
            cv2.COLOR_RGB2GRAY)
    except:
        continue
    test_x.append((black_n_white_img-127.5)/127.5)
    test_y.append((color_img-127.5)/127.5)
```

Note that images are represented with pixel values ranging from 0 to 255. We are normalizing pixel values and bringing them into the range [-1, 1] by subtracting and dividing by 127.5. Data normalization makes the optimization of deep learning models smoother and more stable.

Now that we have successfully prepared our dataset with train, validation, and test set splits, let's check a few samples to confirm that the data has been prepared correctly. The following code block chooses some random training set images and plots them.

Let's plot some input images to get a sense of the data:

```
# input images
indexes = np.random.choice(range(0,4000), size=3)

print("Input Samples (black and white): ")
plt.figure(figsize=(7,7))
for i in range(3):
    plt.subplot(330+1+i)
    plt.imshow((train_x[indexes[i]]+1.0)/2.0, cmap='gray')
    plt.axis('off')
plt.show()
```

We will also be plotting the output versions (colored versions) of these randomly chosen images:

```
# corresponding output images
print("Output Samples (colored): ")
plt.figure(figsize=(7,7))
for i in range(3):
    plt.subplot(330+1+i)
```

```
      plt.imshow((train_y[indexes[i]]+1.0)/2.0)
      plt.axis('off')
  plt.show()
```

If everything is correct, we should see input-output pair images very similar to as in *Figure 7.2*.

Figure 7.2 – Sample input-output pairs for data verification

As we are dealing with image data here, we will be working with a **Convolutional Neural Network (CNN)**-based model so that we can extract useful features from image data. The current research shows that CNNs can be very useful in extracting features and other useful information from image data. As we will be working with CNNs here, we need to convert our image dataset into NumPy arrays, and also add one channel dimension to each black-and-white input image (CNNs accept image input as a three-dimensional array, one dimension each for width, height, and channels). A colored image will already have three channels, one for each color value – R, G, and B. The following code block prepares our final dataset as per the steps described previously:

```
train_x = np.expand_dims(np.array(train_x),-1)
val_x = np.expand_dims(np.array(val_x),-1)
test_x = np.expand_dims(np.array(test_x),-1)

train_y = np.array(train_y)
val_y = np.array(val_y)
test_y = np.array(test_y)
```

Now, once again, let's check the dimensions of our dataset splits:

```
print(train_x.shape, train_y.shape, val_x.shape, \
    val_y.shape, test_x.shape, test_y.shape)
```

It should print something like this:

```
(4430, 80, 80, 1) (4430, 80, 80, 3) (1478, 80, 80, 1) (1478, 80, 80,
3) (1476, 80, 80, 1) (1476, 80, 80, 3)
```

Everything looks great from a data perspective. Let's jump into defining our neural network architecture.

In this experiment, we will define a TensorFlow-based CNN that takes black-and-white images as input and predicts their colored variants as output. The model architecture can be broadly divided into two parts – **encoder** and **decoder**. The encoder part of the model takes a black-and-white image as input and extracts useful features from it by passing it through four down-sampling convolutional layers. Each convolutional layer is followed by layers of **LeakyReLU** activation and **batch normalization** except for the last layer, which has a **dropout** layer in place of batch normalization. After passing through the encoder model, an input image with the dimensions (80, 80, 1) changes into a feature vector with the dimensions (5, 5, 256).

The second part of the model is called the decoder. The decoder part takes the feature vector from the encoder output and converts it back into a colored version of the corresponding input image. The decoder is made up of four transpose-convolutional or up-sampling layers. Each decoder layer is followed by layers of ReLU activation and batch normalization except for the last layer, which has tanh activation and does not have a normalization layer. tanh activation restricts final output vector values into the range [-1,1], which is desired for our output image.

The following code blocks define the TensorFlow model:

```
def tf_model():
    black_n_white_input = tensorflow.keras.layers.Input( \
        shape=(80, 80, 1))

    enc = black_n_white_input
```

The encoder part starts from here, within the same function:

```
    #Encoder part
    enc = tensorflow.keras.layers.Conv2D(32, kernel_size=3, \
        strides=2, padding='same')(enc)
    enc = tensorflow.keras.layers.LeakyReLU(alpha=0.2)(enc)
    enc = tensorflow.keras.layers.BatchNormalization(momentum=0.8)
(enc)

    enc = tensorflow.keras.layers.Conv2D(64, kernel_size=3, \
        strides=2, padding='same')(enc)
```

```
    enc = tensorflow.keras.layers.LeakyReLU(alpha=0.2)(enc)
    enc = tensorflow.keras.layers.BatchNormalization(momentum=0.8)
(enc)

    enc = tensorflow.keras.layers.Conv2D(128, \
        kernel_size=3, strides=2, padding='same')(enc)
    enc = tensorflow.keras.layers.LeakyReLU(alpha=0.2)(enc)
    enc = tensorflow.keras.layers.BatchNormalization(momentum=0.8)
(enc)

    enc = tensorflow.keras.layers.Conv2D(256, \
        kernel_size=1, strides=2, padding='same')(enc)
    enc = tensorflow.keras.layers.LeakyReLU(alpha=0.2)(enc)
    enc = tensorflow.keras.layers.Dropout(0.5)(enc)
```

The decoder part starts from here, within the same function:

```
    #Decoder part
    dec = enc

    dec = tensorflow.keras.layers.Conv2DTranspose(256, \
        kernel_size=3, strides=2, padding='same')(dec)
    dec = tensorflow.keras.layers.Activation('relu')(dec)
    dec = tensorflow.keras.layers.BatchNormalization(momentum=0.8)
(dec)

    dec = tensorflow.keras.layers.Conv2DTranspose(128, \
        kernel_size=3, strides=2, padding='same')(dec)
    dec = tensorflow.keras.layers.Activation('relu')(dec)
    dec = tensorflow.keras.layers.BatchNormalization(momentum=0.8)
(dec)

    dec = tensorflow.keras.layers.Conv2DTranspose(64, \
        kernel_size=3, strides=2, padding='same')(dec)
    dec = tensorflow.keras.layers.Activation('relu')(dec)
    dec = tensorflow.keras.layers.BatchNormalization(momentum=0.8)
(dec)

    dec = tensorflow.keras.layers.Conv2DTranspose(32, \
        kernel_size=3, strides=2, padding='same')(dec)
    dec = tensorflow.keras.layers.Activation('relu')(dec)
    dec = tensorflow.keras.layers.BatchNormalization(momentum=0.8)
(dec)

    dec = tensorflow.keras.layers.Conv2D(3, kernel_size=3,\
        padding='same')(dec)
```

Finally, add tanh activation to get the required output image in colored format:

```
color_image = tensorflow.keras.layers.Activation('tanh')(dec)

return black_n_white_input, color_image
```

Now, let's create a TensorFlow model object and print the summary of our model:

```
black_n_white_input, color_image = tf_model()
model = tensorflow.keras.models.Model( \
    inputs=black_n_white_input, outputs=color_image)
model.summary()
```

This should print the model summary as shown in *Figure 7.3*.

```
Model: "model"

Layer (type)                  Output Shape          Param #
==========================================================
input_1 (InputLayer)          [(None, 80, 80, 1)]   0

conv2d (Conv2D)               (None, 40, 40, 32)    320

leaky_re_lu (LeakyReLU)       (None, 40, 40, 32)    0

batch_normalization (BatchN   (None, 40, 40, 32)    128
ormalization)

conv2d_1 (Conv2D)             (None, 20, 20, 64)    18496

leaky_re_lu_1 (LeakyReLU)     (None, 20, 20, 64)    0

batch_normalization_1 (Batc   (None, 20, 20, 64)    256
hNormalization)

conv2d_2 (Conv2D)             (None, 10, 10, 128)   73856

             ●    ●    ●    ●

             ●    ●    ●    ●

batch_normalization_6 (Batc   (None, 80, 80, 32)    128
hNormalization)

conv2d_4 (Conv2D)             (None, 80, 80, 3)     867

activation_4 (Activation)     (None, 80, 80, 3)     0

==========================================================
Total params: 1,106,755
Trainable params: 1,105,347
Non-trainable params: 1,408
```

Figure 7.3 – TensorFlow model summary (see full summary on GitHub)

As we can see from the summary, our model has roughly 1.1 million trainable parameters. The next step is to compile the TensorFlow model:

```
_optimizer = tensorflow.keras.optimizers.Adam(\
    learning_rate=0.0002, beta_1=0.5)
model.compile(loss='mse', optimizer=_optimizer)
```

We are using the Adam optimizer with a learning rate of 0.0002 and the `beta_1` parameter with a value of 0.5. Here, `beta_1` represents the value for the exponential decay rate for the first-moment estimates and the learning rate tells the optimizer the rate of updating the model parameter values during training. The rest of the parameter values are kept as the default. The idea is to pass a black-and-white image and reconstruct its colored version, so we will be using the **Mean Squared Error (MSE)** loss function as a reconstruction loss on the pixel level.

We are all set to start the training now. We will train our model for about 100 epochs, with a batch size of 128 for this experiment, and check the results. The following code snippet starts the training:

```
history = model.fit(
    train_x,
    train_y,
    batch_size=128,
    epochs=100,
    validation_data=(val_x, val_y),
)
```

The output logs should look something like the following:

```
Epoch 1/100
35/35 [==============================] - 25s 659ms/step - loss: 0.2940
- val_loss: 0.1192
Epoch 2/100
35/35 [==============================] - 20s 585ms/step - loss: 0.1117
- val_loss: 0.0917
Epoch 3/100
35/35 [==============================] - 20s 580ms/step - loss: 0.0929
- val_loss: 0.0784
Epoch 4/100
35/35 [==============================] - 20s 577ms/step - loss: 0.0832
- val_loss: 0.0739
Epoch 5/100
35/35 [==============================] - 20s 573ms/step - loss: 0.0778
- val_loss: 0.0698
.  .  .  .  .
.  .  .  .  .
.  .  .  .  .
.  .  .  .  .
```

```
Epoch 100/100
35/35 [==============================] - 20s 559ms/step - loss: 0.0494
- val_loss: 0.0453
```

To check whether our training went smoothly, we can have a look at the loss charts from the `history` variable:

```
plt.plot(history.history['loss'])
plt.plot(history.history['val_loss'])
plt.title('model loss')
plt.ylabel('loss')
plt.xlabel('epoch')
plt.legend(['train', 'val'], loc='upper right')
plt.show()
```

The preceding snippet will plot the training and validation loss as a line chart for all the training epochs. The output graphs should look something like *Figure 7.4*. As we can see, training and validation loss are consistently decreasing as training progresses. It is reassuring that our training is going in the right direction.

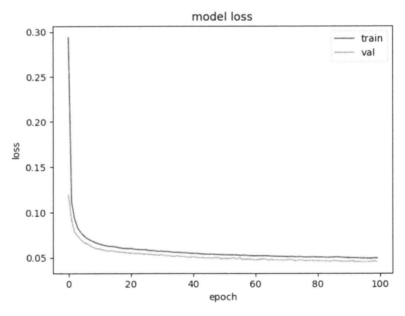

Figure 7.4 – Training and validation loss

The final step is now to check the results on an unseen test dataset. The following code chooses some random samples from `test_set` and generates model outputs for them. We've also plotted input images, model-generated colored images, and actual colored images for understanding purposes:

```
samples = np.random.choice(range(0, len(test_files)), size=5)
```

Plot a few test images to verify the model outputs:

```
# show input images
print("Black-n-White Input images!")
plt.figure(figsize=(8,8))
for i in range(5):
    plt.subplot(550+1+i)
    plt.imshow((test_x[samples[i]]+1.0)/2.0, cmap='gray')
    plt.axis('off')
plt.show()
```

Here, the model generates a colored version:

```
# generate color images from model
print("Model Outputs!")
plt.figure(figsize=(8,8))
for i in range(5):
    plt.subplot(550+1+i)
    model_input = test_x[samples[i]]
    output = model.predict(np.array([model_input]))
    plt.imshow((output[0]+1.0)/2.0)
    plt.axis('off')
plt.show()
```

Also, plot the real colored version for reference:

```
# show real color output images
print("Real Color Version!")
plt.figure(figsize=(8,8))
for i in range(5):
    plt.subplot(550+1+i)
    plt.imshow((test_y[samples[i]]+1.0)/2.0)
    plt.axis('off')
plt.show()
```

When plotting model outputs or input images, we add 1.0 to the image array and divide by 2.0. We are doing this because during data preprocessing, we normalized image pixel values into the range [-1,1]. But ideally, image pixel values can't be negative, so we need to inverse our transformation for image plotting purposes. So, adding 1.0 and dividing by 2.0 brings pixel values into the range [0,1], which is supported by Matplotlib for plotting. See *Figure 7.5*.

Figure 7.5 – Black-and-white to color model output

As we can see from the preceding outputs, our model is learning some kind of colorization, which is, of course, not ideal but still looks pretty good. An interesting thing to notice is that it is not filling the color gradient randomly; we can clearly spot the main objects as they have a different contrast from the background. Given that we had a very small model and small training dataset, this performance is quite promising.

However, this is not the best model architecture for solving the image colorization problem. Nowadays, generative models such as **Generative Adversarial Networks** (**GANs**) provide the best results for such problems. We will study GANs later in this book, but for now, let's stick to this simple experiment. Next, we will work with other Vertex AI tools that will make our lives easier when it comes to experimentation.

# Packaging a model to submit it to Vertex AI as a training job

The previous section demonstrated a small image colorization experiment on a Vertex AI Workbench notebook. Notebooks are great for small-scale and quick experiments, but when it comes to large-scale experiments (with more compute and/or memory requirements), it is advised to launch them as a Vertex AI job and specify desired machine specifications (accelerators such as GPU or TPU if needed) for optimal experimentation. Vertex AI jobs also let us execute tons of experiments in parallel without waiting for the results of a single experiment. Experiment tracking is also quite easy with Vertex AI jobs, so it becomes easier to compare your latest experiments with past experiments with the help of saved metadata and the Vertex AI UI. Now, let's use our model experimentation setup from the previous section and launch it as a Vertex AI training job.

> **Important note**
>
> Vertex AI jobs run in a containerized environment, so in order to launch an experiment, we must package our entire code (including reading data, preprocessing, model building, training, and evaluation) into a single script to be launched within the container. Google Cloud provides tons of prebuilt container images for training and evaluation (with dependencies pre-installed for desired frameworks such as TensorFlow, PyTorch, etc.). Plus, we also have the flexibility of defining our own custom container with any kind of dependencies that we may need.

For the experiment from the previous section, as we downloaded open source data into our Jupyter environment, this data is not yet present in **Google Cloud Storage** (**GCS**) (i.e., a GCS bucket or BigQuery). So, first, we need to store this data somewhere such that our Vertex AI training job can read it from within the training container. To make things easier for us, we will upload our pre-processed data into a storage bucket. This will save us the effort of preparing data again within the job container. We can use the following script to save our prepared data into a GCS bucket:

```
from io import BytesIO
import numpy as np
from tensorflow.python.lib.io import file_io


dest = 'gs://data-bucket-417812395597/' # Destination to save in GCS
## saving training data
np.save(file_io.FileIO(dest+'train_x', 'w'), train_x)
np.save(file_io.FileIO(dest+'train_y', 'w'), train_y)
## saving validation data
np.save(file_io.FileIO(dest+'val_x', 'w'), val_x)
np.save(file_io.FileIO(dest+'val_y', 'w'), val_y)
## saving test data
np.save(file_io.FileIO(dest+'test_x', 'w'), test_x)
np.save(file_io.FileIO(dest+'test_y', 'w'), test_y)
```

Note that before executing this code, we must create a bucket where we want to store these NumPy arrays. In this case, we have already created one bucket with the name `data-bucket-417812395597`.

We can read these NumPy arrays in any number of training jobs/experiments using the following script:

```
train_x = np.load(BytesIO(file_io.read_file_to_string( \
    dest+'train_x', binary_mode=True)))
train_y = np.load(BytesIO(file_io.read_file_to_string( \
    dest+'train_y', binary_mode=True)))
val_x = np.load(BytesIO(file_io.read_file_to_string( \
    dest+'val_x', binary_mode=True)))
val_y = np.load(BytesIO(file_io.read_file_to_string( \
    dest+'val_y', binary_mode=True)))
test_x = np.load(BytesIO(file_io.read_file_to_string( \
    dest+'test_x', binary_mode=True)))
test_y = np.load(BytesIO(file_io.read_file_to_string( \
    dest+'test_y', binary_mode=True)))
```

Our data requirements are now all set. Next, let's work on setting up our Vertex AI training job.

First, we will install some useful packages required to define and launch Vertex AI jobs:

```
# Install the packages
! pip3 install --upgrade google-cloud-aiplatform \
                         google-cloud-storage \
                         pillow
```

Once package installation is done, we will move to a new notebook and import useful libraries:

```
import numpy as np
import glob
import matplotlib.pyplot as plt
import os
from google.cloud import aiplatform
%matplotlib inline
```

Next, we will define our project configurations:

```
PROJECT_ID='41xxxxxxxx7'
REGION='us-west2'
BUCKET_URI='gs://my-training-artifacts'
```

Note that we have created a bucket with the name `my-training-artifacts` to store all the intermediate metadata and artifacts as a result of our Vertex AI job.

Next, let's initialize the Vertex AI SDK with our project configurations:

```
aiplatform.init(project=PROJECT_ID, location=REGION, \
    staging_bucket=BUCKET_URI)
```

For our experimentations, we will be using prebuilt TensorFlow images as our model is also based on TensorFlow. Let's define the images to be used:

```
TRAIN_VERSION = "tf-cpu.2-9"
DEPLOY_VERSION = "tf2-cpu.2-9"
TRAIN_IMAGE = "us-docker.pkg.dev/vertex-ai/training/{}:latest".
format(TRAIN_VERSION)
DEPLOY_IMAGE = "us-docker.pkg.dev/vertex-ai/prediction/{}:latest".
format(DEPLOY_VERSION)
```

In this section, we will just launch a simple training job. In the next sections, we will also deploy and test our trained models.

Next, let's define some command-line arguments for our training (these can be modified on a per-need basis):

```
JOB_NAME = "vertex_custom_training"
MODEL_DIR = "{}/{}".format(BUCKET_URI, JOB_NAME)

TRAIN_STRATEGY = "single"
EPOCHS = 20
STEPS = 100

CMDARGS = [
    "--epochs=" + str(EPOCHS),
    "--steps=" + str(STEPS),
    "--distribute=" + TRAIN_STRATEGY,
]
```

We should also provide a meaningful job name; it will help us distinguish our experiment from other experiments running in parallel.

The next step is to write down our entire training script – starting from reading data, defining the model, training, and saving the model into a single file. We will write down our entire code from the previous section into a file with the name task.py. The following are the contents of our task. py file:

```
%%writefile task.py
# Single, Mirror and Multi-Machine Distributed Training

import tensorflow as tf
```

```
import tensorflow
from tensorflow.python.client import device_lib
import argparse
import os
import sys
from io import BytesIO
import numpy as np
from tensorflow.python.lib.io import file_io
```

The following part of the file parses command-line arguments:

```
# parse required arguments
parser = argparse.ArgumentParser()
parser.add_argument('--lr', dest='lr', \
                    default=0.001, type=float, \
                    help='Learning rate.')
parser.add_argument('--epochs', dest='epochs', \
                    default=10, type=int, \
                    help='Number of epochs.')
parser.add_argument('--steps', dest='steps', \
                    default=35, type=int, \
                    help='Number of steps per epoch.')
parser.add_argument('--distribute', dest='distribute', \
                    type=str, default='single', \
                    help='distributed training strategy')
args = parser.parse_args()
```

Here, we print some version and environment configurations to keep track of current settings:

```
print('Python Version = {}'.format(sys.version))
print('TensorFlow Version = {}'.format(tf.__version__))
print('TF_CONFIG = {}'.format(os.environ.get('TF_CONFIG', \
    'Not found')))
print('DEVICES', device_lib.list_local_devices())
```

Here, we define a training strategy:

```
# Single Machine, single compute device
if args.distribute == 'single':
    if tf.test.is_gpu_available():
        strategy = tf.distribute.OneDeviceStrategy(device="/gpu:0")
    else:
        strategy = tf.distribute.OneDeviceStrategy(device="/cpu:0")
# Single Machine, multiple compute device
elif args.distribute == 'mirror':
```

```
    strategy = tf.distribute.MirroredStrategy()
# Multiple Machine, multiple compute device
elif args.distribute == 'multi':
    strategy = tf.distribute.experimental.
MultiWorkerMirroredStrategy()

# Multi-worker configuration
print('num_replicas_in_sync = {}'.format(strategy.num_replicas_in_
sync))
```

Now, we prepare the dataset for training, validation, and testing:

```
# Preparing dataset
BUFFER_SIZE = 10000
BATCH_SIZE = 128

def make_datasets_unbatched():
    # Load train, validation and test sets
    dest = 'gs://data-bucket-417812395597/'
    train_x = np.load(BytesIO(
        file_io.read_file_to_string(dest+'train_x', \
            binary_mode=True)
    ))
    train_y = np.load(BytesIO(
        file_io.read_file_to_string(dest+'train_y', \
            binary_mode=True)
    ))
    val_x = np.load(BytesIO(
        file_io.read_file_to_string(dest+'val_x', \
            binary_mode=True)
    ))
    val_y = np.load(BytesIO(
        file_io.read_file_to_string(dest+'val_y', \
            binary_mode=True)
    ))
    test_x = np.load(BytesIO(
        file_io.read_file_to_string(dest+'test_x', \
            binary_mode=True)
    ))
    test_y = np.load(BytesIO(
        file_io.read_file_to_string(dest+'test_y', \
            binary_mode=True)
    ))
    return train_x, train_y, val_x, val_y, test_x, test_y
```

Now, we define our TensorFlow model as discussed before:

```
def tf_model():
    black_n_white_input = tensorflow.keras.layers.Input(shape=(80, 80,
1))

    enc = black_n_white_input
```

Here is the definition of Encoder part of the TF model:

```
    #Encoder part
    enc = tensorflow.keras.layers.Conv2D(
        32, kernel_size=3, strides=2, padding='same'
    )(enc)
    enc = tensorflow.keras.layers.LeakyReLU(alpha=0.2)(enc)
    enc = tensorflow.keras.layers.BatchNormalization(momentum=0.8)
(enc)

    enc = tensorflow.keras.layers.Conv2D(
        64, kernel_size=3, strides=2, padding='same'
    )(enc)
    enc = tensorflow.keras.layers.LeakyReLU(alpha=0.2)(enc)
    enc = tensorflow.keras.layers.BatchNormalization(momentum=0.8)
(enc)

    enc = tensorflow.keras.layers.Conv2D(
        128, kernel_size=3, strides=2, padding='same'
    )(enc)
    enc = tensorflow.keras.layers.LeakyReLU(alpha=0.2)(enc)
    enc = tensorflow.keras.layers.BatchNormalization(momentum=0.8)
(enc)

    enc = tensorflow.keras.layers.Conv2D(
        256, kernel_size=1, strides=2, padding='same'
    )(enc)
    enc = tensorflow.keras.layers.LeakyReLU(alpha=0.2)(enc)
    enc = tensorflow.keras.layers.Dropout(0.5)(enc)
```

The Encoder part is now done. Next we define the decoder part of the model within the same function:

```
    #Decoder part
    dec = enc

    dec = tensorflow.keras.layers.Conv2DTranspose(
        256, kernel_size=3, strides=2, padding='same'
    )(dec)
```

```
    dec = tensorflow.keras.layers.Activation('relu')(dec)
    dec = tensorflow.keras.layers.BatchNormalization(momentum=0.8)
(dec)

    dec = tensorflow.keras.layers.Conv2DTranspose(
        128, kernel_size=3, strides=2, padding='same'
    )(dec)
    dec = tensorflow.keras.layers.Activation('relu')(dec)
    dec = tensorflow.keras.layers.BatchNormalization(momentum=0.8)
(dec)

    dec = tensorflow.keras.layers.Conv2DTranspose(
        64, kernel_size=3, strides=2, padding='same'
    )(dec)
    dec = tensorflow.keras.layers.Activation('relu')(dec)
    dec = tensorflow.keras.layers.BatchNormalization(momentum=0.8)
(dec)

    dec = tensorflow.keras.layers.Conv2DTranspose(
        32, kernel_size=3, strides=2, padding='same'
    )(dec)
    dec = tensorflow.keras.layers.Activation('relu')(dec)
    dec = tensorflow.keras.layers.BatchNormalization(momentum=0.8)
(dec)

    dec = tensorflow.keras.layers.Conv2D(
        3, kernel_size=3, padding='same'
    )(dec)
```

Here, we apply tanh activation function to get the colored output image -

```
    color_image = tensorflow.keras.layers.Activation('tanh')(dec)

    return black_n_white_input, color_image
```

Now, we are ready to build and compile our TensorFlow model:

```
# Build the and compile TF model
def build_and_compile_tf_model():
    black_n_white_input, color_image = tf_model()
    model = tensorflow.keras.models.Model(
        inputs=black_n_white_input,
        outputs=color_image
    )
    _optimizer = tensorflow.keras.optimizers.Adam(
        learning_rate=0.0002,
```

```
        beta_1=0.5
    )
    model.compile(
        loss='mse',
        optimizer=_optimizer
    )
    return model
```

The following block launches training with the defined settings and saves the trained model:

```
# Train the model
NUM_WORKERS = strategy.num_replicas_in_sync
# Here the batch size scales up by number of workers since
# `tf.data.Dataset.batch` expects the global batch size.
GLOBAL_BATCH_SIZE = BATCH_SIZE * NUM_WORKERS
MODEL_DIR = os.getenv("AIP_MODEL_DIR")

train_x, train_y, _, _, _, _ = make_datasets_unbatched()

with strategy.scope():
    # Creation of dataset, and model building/compiling need to be
within
    # `strategy.scope()`.
    model = build_and_compile_tf_model()

model.fit(
    train_x,
    train_y,
    epochs=args.epochs,
    steps_per_epoch=args.steps
)
model.save(MODEL_DIR)
```

Now that we have all the configurations set up and our training script, task.py, is ready, we are all set to define and launch our custom training job on Vertex AI.

Let's define our custom Vertex AI training job:

```
job = aiplatform.CustomTrainingJob(
    display_name=JOB_NAME,
    script_path="task.py",
    container_uri=TRAIN_IMAGE,
    requirements=[],
    model_serving_container_image_uri=DEPLOY_IMAGE,
)
```

The final step is to launch the job:

```
MODEL_DISPLAY_NAME = "tf_bnw_to_color"
# Start the training job
model = job.run(
    model_display_name=MODEL_DISPLAY_NAME,
    args=CMDARGS,
    machine_type = "n1-standard-16",
    replica_count=1,
)
```

This setup launches a Vertex AI custom training job on an n1-standard-16 machine as defined as a parameter in the preceding job.run method. When we launch the job in a notebook cell, it gives us a URL to the Google Cloud console UI. By clicking on it, we can monitor our job logs within the Vertex AI UI.

A Vertex AI training job looks something like *Figure 7.6* in the Google Cloud console UI. Here, we can re-verify the configurations and parameters of our job that we had defined at the time of launch:

Figure 7.6 – Vertex AI training job

The Vertex AI UI lets us monitor near real-time logs of all the training/custom jobs. We can monitor our training within the UI, and it looks something like *Figure 7.7*:

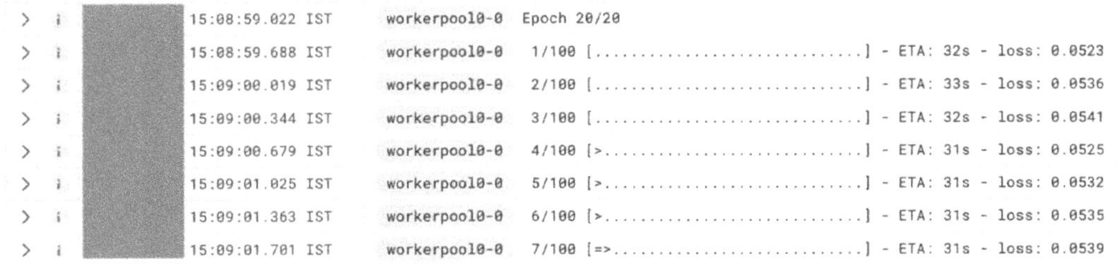

```
>  i         15:08:59.022 IST    workerpool0-0  Epoch 20/20
>  i         15:08:59.688 IST    workerpool0-0  1/100 [..............................] - ETA: 32s - loss: 0.0523
>  i         15:09:00.019 IST    workerpool0-0  2/100 [..............................] - ETA: 33s - loss: 0.0536
>  i         15:09:00.344 IST    workerpool0-0  3/100 [..............................] - ETA: 32s - loss: 0.0541
>  i         15:09:00.679 IST    workerpool0-0  4/100 [>.............................] - ETA: 31s - loss: 0.0525
>  i         15:09:01.025 IST    workerpool0-0  5/100 [>.............................] - ETA: 31s - loss: 0.0532
>  i         15:09:01.363 IST    workerpool0-0  6/100 [>.............................] - ETA: 31s - loss: 0.0535
>  i         15:09:01.701 IST    workerpool0-0  7/100 [=>............................] - ETA: 31s - loss: 0.0539
```

Figure 7.7 – Real-time logs for Vertex AI training job on Google Cloud console

Going through the logs may not be the best way to monitor training progress as we may want to track a few parameters, such as loss and accuracy. In the next section, we will learn about how to set up TensorBoard-based live monitoring of training progress. However, these logs can be really handy for debugging purposes; if our pipeline fails in between before completing the execution successfully, we can always check these logs to identify the root cause.

## Monitoring model training progress

In the previous section, we saw how easy it is to launch a Vertex AI custom training job with desired configurations and machine types. These Vertex AI training jobs are really useful for running large-scale experiments where training uses high compute (multiple GPUs or TPUs) and also may run for a few days. Such long-running experiments are not very feasible to run in a Jupyter Notebook-based environment. Another great thing about launching Vertex AI jobs is that all the metadata and lineage are tracked in a systematic way so that we can come back later and look into our past experiments and compare them with the latest ones in an easy and accurate way.

Another important aspect is monitoring the live progress of training jobs (including metrics such as loss and accuracy). For this purpose, we can easily set up Vertex AI TensorBoard within our Vertex AI job and track the progress in a near real-time fashion. In this section, we will set up a TensorBoard instance for our previous experiment.

Most of the code/scripts will be similar to the previous section. Here, we will just examine the modifications needed to set up TensorBoard monitoring.

Firstly, we need to make small changes in the task.py file to account for TensorFlow callbacks as we want to monitor training loss. To keep things clean, we will modify a copy of the task.py file that we have renamed to task2.py. The following are the changes in the model.fit function:

```
### Create a TensorBoard callback and write to the gcs path provided
by AIP_TENSORBOARD_LOG_DIR
tensorboard_callback = tf.keras.callbacks.TensorBoard(
    log_dir=os.environ['AIP_TENSORBOARD_LOG_DIR'],
```

```
    histogram_freq=1)

model.fit(
    train_x,
    train_y,
    epochs=args.epochs,
    steps_per_epoch=args.steps,
    callbacks=[tensorboard_callback],
)
```

In the preceding script, we have just defined a TensorFlow callback object and also passed it into the `model.fit` function.

Working with TensorBoard requires a service account to be in place (instead of individual user accounts). If we already don't have a service account set up, we can use the following script to quickly set up a service account. A service account is used to grant permissions to services, VMs, and other tooling on Google Cloud:

```
SERVICE_ACCOUNT="dummy-sa"
IS_COLAB=False
if (
    SERVICE_ACCOUNT == ""
    or SERVICE_ACCOUNT is None
    or SERVICE_ACCOUNT == "dummy-sa"
):
    # Get your service account from gcloud
    if not IS_COLAB:
        shell_output = ! gcloud auth list 2>/dev/null
        SERVICE_ACCOUNT = shell_output[2].replace("*", \
            "").strip()
```

If we are working with colab, the following code snippet will create a service account accordingly:

```
    else:  # IS_COLAB:
        shell_output = ! gcloud projects describe  $PROJECT_ID
        project_number = shell_output[-1].split(":")[1].strip().
replace("'", "")
        SERVICE_ACCOUNT = f"{project_number}-compute@developer.
gserviceaccount.com"

    print("Service Account:", SERVICE_ACCOUNT)
```

The next step is to create a Vertex AI TensorBoard instance that we will use for monitoring our training.

Set up the TensorBoard instance:

```
TENSORBOARD_NAME = "training-monitoring"  # @param {type:"string"}

if (
    TENSORBOARD_NAME == ""
    or TENSORBOARD_NAME is None
    or TENSORBOARD_NAME == "training-monitoring"
):
    TENSORBOARD_NAME = PROJECT_ID + "-tb-" + TIMESTAMP

tensorboard = aiplatform.Tensorboard.create(
    display_name=TENSORBOARD_NAME, project=PROJECT_ID, \
        location=REGION
)
Let's verify if the TensorBoard instance was successfully created or
not - TENSORBOARD_RESOURCE_NAME = tensorboard.gca_resource.name
print("TensorBoard resource name:", TENSORBOARD_RESOURCE_NAME)
```

We need a staging bucket for our Vertex AI job so that it can write event logs into that location:

```
BUCKET_URI = "gs://tensorboard-staging"  # @param {type:"string"}

if BUCKET_URI == "" or BUCKET_URI is None or BUCKET_URI == "gs://
[your-bucket-name]":
    BUCKET_URI = "gs://" + PROJECT_ID + "aip-" + TIMESTAMP

! gsutil mb -l {REGION} -p {PROJECT_ID} {BUCKET_URI}

GCS_BUCKET_OUTPUT = BUCKET_URI + "/output/"
```

We are now all set to define our custom training job:

```
JOB_NAME = "tensorboard-example-job-{}".format(TIMESTAMP)
BASE_OUTPUT_DIR = "{}{}".format(GCS_BUCKET_OUTPUT, JOB_NAME)

job = aiplatform.CustomTrainingJob(
    display_name=JOB_NAME,
    script_path="task2.py",
    container_uri=TRAIN_IMAGE,
    requirements=[],
    model_serving_container_image_uri=DEPLOY_IMAGE,
    staging_bucket=BASE_OUTPUT_DIR,
)
```

We can now launch the Vertex AI job using the following script. Here, we can choose the machine type and also specify the `replica_count` parameter, which controls the number of replicas to run for the current job:

```
MODEL_DISPLAY_NAME = "tf_bnw_to_color_tb"
# Start the training job
model = job.run(
    model_display_name=MODEL_DISPLAY_NAME,
    service_account=SERVICE_ACCOUNT,
    tensorboard=TENSORBOARD_RESOURCE_NAME,
    args=CMDARGS,
    machine_type = "n1-standard-8",
    replica_count=1,
)
```

Once we launch the job, it will give us the URL for locating the Vertex AI job in the Google Cloud console UI like in the previous section; but this time, it will also give us a URL to the Vertex TensorBoard UI. Using this URL, we will be able to monitor our training in a near real-time fashion.

This is how it looks for our little experiment (see *Figure 7.8*):

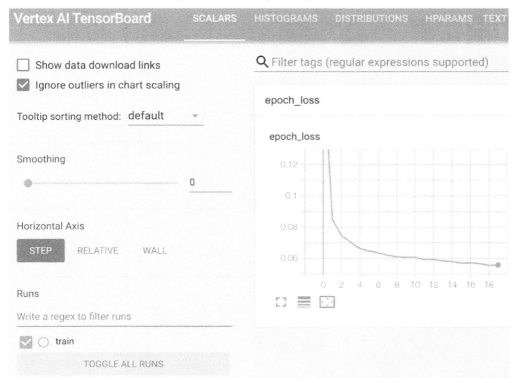

Figure 7.8 – Vertex TensorBoard for real-time monitoring of experiments

We can configure it to show more desired metrics for our experiments. Now that we are able to launch Vertex AI training, monitor it, and also save our TensorFlow-trained model, let's move on to the model evaluation part.

# Evaluating trained models

In this section, we will take the already trained model from the previous section and launch a batch inference job on the test data. The first step here will be to load our test data into a Jupyter Notebook:

```
from io import BytesIO
import numpy as np
from tensorflow.python.lib.io import file_io

dest = 'gs://data-bucket-417812395597/'
test_x = np.load(BytesIO(file_io.read_file_to_string(dest+'test_x',\
    binary_mode=True)))
test_y = np.load(BytesIO(file_io.read_file_to_string(dest+'test_y',\
    binary_mode=True)))
print(test_x.shape, test_y.shape)
```

The next step is to create a JSON payload of instances from our test data and save it in a cloud storage location. The batch inference module will be able to read these instances and perform inference:

```
import json

BATCH_PREDICTION_INSTANCES_FILE = "batch_prediction_instances.jsonl"

BATCH_PREDICTION_GCS_SOURCE = (
    BUCKET_URI + "/batch_prediction_instances/" + BATCH_PREDICTION_
INSTANCES_FILE
)
```

Here we convert the input images to a serializable format so that the prediction service can accept input as a JSON file:

```
# converting to serializable format
x_test = [(image).astype(np.float32).tolist() for image in test_x]

# Write instances at JSONL
with open(BATCH_PREDICTION_INSTANCES_FILE, "w") as f:
    for x in x_test:
        f.write(json.dumps(x) + "\n")

# Upload to Cloud Storage bucket
```

```
! gsutil cp batch_prediction_instances.jsonl BATCH_PREDICTION_GCS_
SOURCE

print("Uploaded instances to: ", BATCH_PREDICTION_GCS_SOURCE)
```

Now our test dataset instances are ready in a cloud storage bucket. We can launch batch prediction over them, and the batch inference module will save the output results into a new folder inside the same bucket:

```
MIN_NODES = 1
MAX_NODES = 1

# The name of the job
BATCH_PREDICTION_JOB_NAME = "bnw_to_color_batch_prediction"

# Folder in the bucket to write results to
DESTINATION_FOLDER = "batch_prediction_results"

# The Cloud Storage bucket to upload results to
BATCH_PREDICTION_GCS_DEST_PREFIX = BUCKET_URI + "/" + DESTINATION_
FOLDER
```

Here, we call the batch prediction service using the SDK:

```
# Make SDK batch_predict method call
batch_prediction_job = model.batch_predict(
    instances_format="jsonl",
    predictions_format="jsonl",
    job_display_name=BATCH_PREDICTION_JOB_NAME,
    gcs_source=BATCH_PREDICTION_GCS_SOURCE,
    gcs_destination_prefix = BATCH_PREDICTION_GCS_DEST_PREFIX,
    model_parameters=None,
    starting_replica_count=MIN_NODES,
    max_replica_count=MAX_NODES,
    machine_type="n1-standard-4",
    sync=True,
)
```

We can also monitor the progress of the batch prediction job within the Google Cloud console UI if needed. Once this job finishes, we can check the outputs inside the defined destination folder.

# Summary

In the chapter, we learned how to work with a Vertex AI-based managed training environment and launch custom training jobs. Launching custom training jobs on Vertex AI comes with a number of advantages, such as managed metadata tracking, no need to actively monitor jobs, and the ability to launch any number of experiments in parallel, choose your desired machine specifications to run your experiments, monitor training progress and results in near-real time fashion using the Cloud console UI, and run managed batch inference jobs on a saved model. It is also tighly integrated with other GCP products.

After reading this chapter, you should be able to develop and run custom deep learning models (using frameworks such as TensorFlow) on Vertex AI Workbench notebooks. Secondly, you should be able to launch long-running Vertex AI custom training jobs and also understand the advantages of the managed Vertex AI training framework. The managed Google Cloud console interface and TensorBoard make it easy to monitor and evaluate various Vertex AI training jobs.

Now that we have a good understanding of training models using Vertex AI on GCP, we will learn about model explainability in the next chapter.

# 8

# ML Model Explainability

In the rapidly evolving world of **machine learning** (**ML**) and **artificial intelligence** (**AI**), developing models capable of delivering accurate predictions is no longer the sole objective. As organizations increasingly rely on data-driven decision-making, understanding the rationale behind a model's predictions becomes paramount. The growing need for explainability in ML models stems from ethical, regulatory, and practical concerns, and it is here that the concept of **Explainable AI** (**XAI**) comes into play.

This chapter delves into the intricacies of Explainable ML models, a critical component in the MLOps landscape, with a focus on their implementation in the Google Cloud ecosystem. Although a comprehensive exploration of XAI techniques and tools is beyond this chapter's scope, we aim to equip you with the knowledge and skills to build transparent, interpretable, and accountable ML models using the Explainable ML tools available on GCP.

The following topics will be covered in this chapter:

- What is Explainable AI and why is it important for MLOps practitioners?
- Overview of Explainable AI techniques
- Explainable AI features available in Google Cloud Vertex AI
- Hands-on exercises for using Vertex AI's explainability features

As we journey through this chapter, we will establish the importance of explainability and its role in enhancing trust, accountability, and fairness in ML models. Next, we will discuss various techniques for achieving explainability in ML, ranging from traditional interpretable models to explanation techniques for more complex models such as deep learning. We will then dive into Google Cloud's XAI offerings, which facilitate the development and evaluation of Explainable ML models.

In addition to providing an understanding of Explainable ML models, this chapter will guide you through practical, hands-on examples, illustrating the application of these concepts in real-world scenarios.

By the end of this chapter, you will be well-equipped to design, deploy, and evaluate Explainable ML models on Google Cloud, ensuring that your organization stays ahead in the race toward ethical and responsible AI adoption.

# What is Explainable AI and why is it important for MLOps practitioners?

XAI refers to methods and techniques that are used in the domain of AI that aim to make the decision-making processes of AI models transparent, interpretable, and understandable to humans. Instead of acting as black boxes where input data goes in and a decision or prediction comes out without clarity on how the decision was reached, XAI seeks to provide insights into the inner workings of models. This transparency allows users, developers, and stakeholders to trust and validate the system's decisions, ensuring they align with ethical, legal, and practical considerations.

As ML continues to advance and its applications permeate various industries, the need for transparent and interpretable models has become a pressing concern. XAI aims to address this by developing techniques for understanding, interpreting, and explaining ML models. For MLOps practitioners working with Google Cloud, incorporating XAI into their workflows can lead to several benefits, including improved trust, regulatory compliance, and enhanced model performance.

First, let's look at the key reasons for the importance of XAI for MLOps practitioners and its impact on the development and deployment of ML models.

## Building trust and confidence

XAI can help MLOps practitioners build trust in their models by providing clear and understandable explanations of how these models make decisions. This is particularly important when dealing with stakeholders who may not have a technical background as the ability to explain model behavior can lead to increased confidence in its predictions. Additionally, having a deeper understanding of the inner workings of a model allows practitioners to better communicate the limitations and strengths of their solutions, which, in turn, can foster trust among collaborators and end users.

### Regulatory compliance

As ML models become more widely adopted, regulatory bodies around the world are increasingly calling for more transparency and accountability in AI systems. XAI techniques can help MLOps practitioners ensure compliance with these regulations by providing insights into the decision-making process of their models. This can be especially important in industries such as healthcare, finance, and human resources, where the consequences of biased or unfair decisions can be significant. By incorporating XAI into their workflows, practitioners can demonstrate that their models adhere to relevant laws and ethical guidelines.

## *Model debugging and improvement*

XAI can be invaluable to MLOps practitioners during the model development and debugging process. By providing insights into how a model is making its predictions, XAI can help identify areas where the model may be underperforming, overfitting, or exhibiting bias. With this information, practitioners can make targeted adjustments to their models, leading to improved performance and more robust solutions. This iterative process can save both time and resources, allowing practitioners to focus on addressing the most critical issues impacting their models.

## *Ethical considerations*

As the power and influence of ML models grow, so too does the responsibility of MLOps practitioners to ensure that these models are used ethically. XAI can help practitioners identify and address any unintended consequences or biases that may arise from their models. By understanding how a model makes its decisions, practitioners can better ensure that their solutions are fair, unbiased, and aligned with ethical principles.

Incorporating XAI into the MLOps workflow within the Google Cloud ecosystem can lead to numerous benefits for practitioners. From building trust with stakeholders and ensuring regulatory compliance to improving model performance and addressing ethical concerns, the importance of XAI cannot be understated. As the field of ML continues to evolve, integrating XAI into MLOps practices will become increasingly essential for the development and deployment of transparent, interpretable, and responsible AI solutions.

# Explainable AI techniques

Different techniques are available to cater to various types of data, including tabular, image, and text data. Each data type presents its own set of challenges and complexities, requiring tailored methods to provide meaningful insights into the decision-making processes of ML models. This subsection will list various XAI techniques applicable to tabular, image, and text data. The following section will dive into the ones available as out-of-the-box features in Google Cloud.

## Global versus local explainability

Explainability can be categorized into two categories: local and global explainability. These terms are sometimes referred to as local and global feature importance:

- **Global explainability** focuses on the overall impact of a feature on the model. This is usually obtained by calculating the average feature attribution values over the entire dataset. A feature with a high absolute value indicates that it significantly influenced the model's predictions.

- **Local explainability** provides insight into how much each feature contributed to the prediction for a specific instance. The feature attribution values give information about the effect of a particular feature on the prediction relative to a baseline prediction.

## Techniques for image data

XAI techniques for image data often focus on visualizing the areas within an image that contribute most significantly to the model's predictions. Here are some key techniques:

- **Integrated Gradients**

  Integrated Gradients is an attribution technique specifically designed for deep learning models, such as neural networks. It computes the gradients of the model's output concerning the input features (pixels in the case of image data) and integrates these gradients over a straight path from a baseline input to the instance of interest. This process assigns an importance value to each pixel in the image, reflecting its contribution to the model's prediction. Integrated Gradients provide insights into the importance of each pixel and can help identify potential biases or shortcomings in the model's predictions:

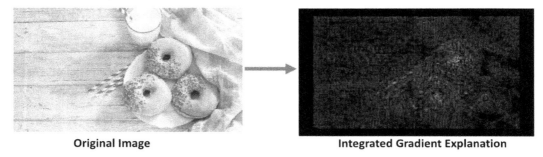

**Original Image**                                    **Integrated Gradient Explanation**

Figure 8.1 – Integrated Gradients explanation

The preceding figure showcases the Integrated Gradients approach to image explanation, which highlights pixels in the image that the model gives high importance to during prediction.

- **eXtended Relevance-weighted Attribution of Importance (XRAI)**

  XRAI is an XAI method for visualizing the most important regions in an image for a given model's prediction. It is an extension of the Integrated Gradients method, which combines pixel-wise attributions with segmentation techniques to generate more coherent and interpretable visualizations. By identifying the most important segments within an image, XRAI provides insights into the model's decision-making process and can help identify potential biases or issues in the model's predictions:

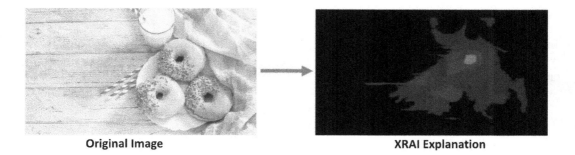

Original Image                                          XRAI Explanation

Figure 8.2 – XRAI

The preceding figure showcases the XRAI approach to an image explanation, which highlights areas of the image that the model gives high importance during prediction.

- **Local Interpretable Model-agnostic Explanations (LIME)**

    LIME is an XAI technique that provides local explanations for individual predictions of any classifier. In the context of image data, LIME generates synthetic data points (perturbed images) around a specific instance, obtains predictions from the model, and fits an interpretable model (for example, linear regression) to these data points, weighted by their proximity to the instance. The resulting model provides insights into the most important regions that contribute to the prediction for the specific instance. By visualizing these regions, practitioners can better understand the model's decision-making process and identify potential biases or issues in the model's predictions.

- **Gradient-weighted Class Activation Mapping (Grad-CAM)**

    Grad-CAM is a visualization technique for deep learning models, specifically **convolutional neural networks** (**CNNs**). It generates heatmap-like visualizations of the most important regions in an image for a given model's prediction. Grad-CAM computes the gradients of the predicted class score concerning the feature maps of the last convolutional layer and then uses these gradients to compute a weighted sum of the feature maps. The resulting heatmap highlights the regions in the image that contribute most to the model's prediction. Grad-CAM provides insights into the model's decision-making process and can help identify potential biases or shortcomings in the model's predictions.

These techniques provide insights into the model's decision-making process and can help identify potential biases or shortcomings in the model's predictions.

## Techniques for tabular data

Tabular data, which is composed of structured rows and columns, is one of the most common data types encountered in ML. Various XAI techniques can be employed to interpret models trained on tabular data:

- **Local Interpretable Model-agnostic Explanations (LIME)**

  As its name suggests, LIME is an XAI technique that provides *local* explanations for individual predictions of any classifier. It does so by approximating the complex model with a simpler, interpretable model (for example, linear regression) within the vicinity of a specific instance. LIME generates synthetic data points around the instance, obtains predictions from the complex model, and fits an interpretable model to these data points, weighted by their proximity to the instance. The resulting model provides insights into the most important features contributing to the prediction for the specific instance.

  In the following figure, we're using a LIME report to explain the decision of an ML model trained to predict the survival probabilities of Titanic's passengers based on their attributes, such as their gender, the fare they paid, the passenger class they were traveling in, and so on. The Titanic survival dataset is a common publically available dataset that's used as an example for classification models. Let's see if LIME can help us get some insights into the model's behavior:

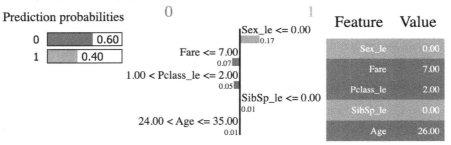

Figure 8.3 – Explaining classification models with LIME

The leftmost chart shows the predicted survival probability for the chosen passenger. In this example, the model predicts that this passenger had a 40% probability of survival based on their key attributes. The chart in the middle shows us the ranked feature importance list generated by using LIME. Based on this chart, it seems that the top three most important features that dictated whether a Titanic passenger survived or not were their sex/gender, fare paid, and the passenger class they were traveling in. This makes sense because we know women were evacuated first, giving them a higher overall chance of survival. We also know that passengers who paid lower fares and had lower-class tickets had their cabins on the lower decks/floors of the ship, which got flooded first and the passengers with higher fares had their cabins on the upper decks, giving them a better chance of survival. So, you can see how LIME and similar techniques can help decipher black box ML models and help us better understand why a particular prediction was made. You will also be glad to know that the Titanic passenger we used for the preceding example was a 26-year-old lady named Miss.

Laina Heikkinen paid $7.925 for a ticket in the third-class passenger section and despite our model giving her a less than 40% chance of survival, she survived.

- **Shapley Additive exPlanations (SHAP)** (native support on Vertex AI)

SHAP is a unified measure of feature importance based on cooperative game theory, providing a consistent and fair way to allocate feature importance. By calculating the Shapley value for each feature, SHAP assigns an importance value that reflects its contribution to the prediction for a specific instance. The Shapley value is calculated by averaging the marginal contributions of a feature across all possible feature combinations. SHAP values provide insights into the most influential features driving a model's predictions and can be used with a variety of models. Recommended model types include non-differentiable models such as ensembles of trees. They can also be used for neural networks, where SHAP can provide insights into the contribution of each input feature to the final prediction made by the network. By analyzing the SHAP values, you can determine which features have the most impact on the network's output and understand the relationship between the inputs and outputs:

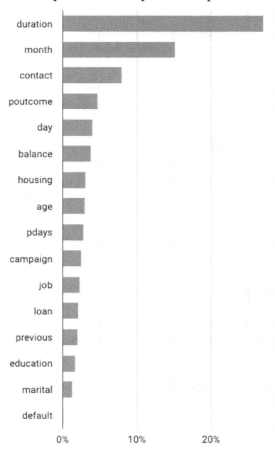

Figure 8.4 – Feature importance graph in Vertex AI based on the Shapley method

- **Permutation feature importance**

    Permutation feature importance is a model-agnostic technique that estimates the importance of each feature by measuring the change in the model's performance when the feature's values are randomly shuffled. This process is repeated several times, and the average decrease in performance is used as an estimate of the feature's importance. By disrupting the relationship between the feature and the target variable, permutation importance helps identify the features that have the greatest impact on the model's predictions.

- **Partial dependence plots (PDP)**

    PDP is a visualization technique that depicts the relationship between a specific feature and the model's predicted outcome while holding all other features constant. By illustrating how a single feature influences the prediction, PDP can help practitioners better understand the behavior of their models and identify potential biases or inconsistencies.

- **Feature importance** (for example, GINI importance and coefficients in linear models)

    Feature importance is a set of techniques that quantify the impact of input features on the model's predictions. These methods can help practitioners identify the most relevant features, enabling them to focus on the most significant variables during model development and debugging. Some common approaches to feature importance are as follows:

    - **GINI Importance**: Used in decision trees and random forests, GINI importance measures the average reduction in impurity (**GINI index**) attributable to a specific feature across all the trees in the forest.

    - **Coefficients in linear models**: In linear regression and logistic regression, the coefficients of the model can be used as a measure of feature importance, indicating the magnitude and direction of the relationship between each feature and the target variable. Larger absolute coefficient values indicate a stronger relationship between the feature and the target variable.

These techniques help practitioners understand the relationships between input features and model predictions, identify the most influential features, and assess the impact of specific features on individual predictions.

## Techniques for text data

In the case of natural language processing models that use text data, the aim is to identify the most significant words or phrases that contribute to the model's predictions. Here are some XAI techniques for text data:

- **Text-specific LIME**: This is an adaptation of LIME specifically designed for text data that provides explanations for individual predictions by highlighting the most important words or phrases.

In the following example, we're using LIME to explain why the ML model we built to classify movie reviews as positive or negative arrives at a particular conclusion:

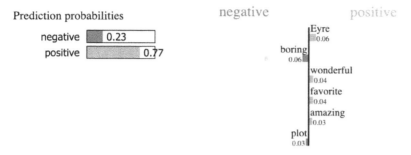

## Text with highlighted words

I had no idea what Jane Eyre was before I saw this miniseries. I had read and watched many classics before, and I believed that most classics were boring, over-worded, and overrated stories with moderately interesting plots at best. This Jane Eyre miniseries completely changed my conceptions.lbr /llbr /lZelah Clarke is a fabulous actress, and she gives a wonderful portrayal of Jane Eyre. Her accent is delightful and her quiet, yet firm nature matches the young governess' character exactly. Timothy Dalton is an amazing Rochester. His passion and energy in the film makes me believe that he was born to play the brooding master of Thornfield Hall. I couldn't sleep at all the night after I had watched this miniseries. The plot is both haunting and inspiring. The characters are masterfully performed, and the story is incredible. This is the best version of Jane Eyre to ever appear on film.lbr /llbr /lI read the book later and was amazed at how closely this miniseries followed Charolette Bronte's writing. Jane Eyre is now my favorite film and book. If you want to see a masterpiece that will change your life, watch the 1983 BBC version of Jane Eyre.

Figure 8.5 – LIME-based explanation for text classification

As we can see, the predicted probability of the movie review being positive is 0.77 (77%). The words highlighted in orange contributed toward the positive probability, while the words highlighted in blue had a significant contribution toward moving the final prediction toward the "negative" label. The chart at the top right shows the respective contribution of each of the highlighted words toward the final decision. For example, if we remove the word "amazing" from the review text, the positive probability will go down by 0.03:

- **Text-specific SHAP**: This is an adaptation of SHAP tailored for text data, attributing importance values to individual words or phrases within a given text

- **Attention mechanisms**: In deep learning models such as Transformers, attention mechanisms can provide insights into the relationships between words and the model's predictions by visualizing the attention weights

Now that we have familiarized ourselves with the various popular XAI techniques, let's look at the different features available within Google Cloud Vertex AI that can help us build XAI solutions using these techniques.

# Explainable AI features available in Google Cloud Vertex AI

**Google Cloud Vertex AI** offers a suite of tools and options tailored to make AI systems more understandable. This section delves into the various XAI options available in Vertex AI, showcasing how this platform is advancing the frontier of transparent ML.

Broadly, the XAI options available in Vertex AI can be divided into two types:

- **Feature-based**: Feature attributions refer to the degree to which each feature in a model contributes to the predictions for a specific instance. When making a prediction request, you receive the predicted values that are generated by your model. However, when requesting an explanation, you receive not only the predictions but also the feature attribution information.

It is important to note that feature attributions are primarily applicable to tabular data but also include built-in visualization capabilities for image data. This makes it easier to understand and interpret the attributions more intuitively.

- **Example-based**: Vertex AI utilizes nearest neighbor search to provide example-based explanations. This method involves finding the closest examples (usually from the training data) to the input and returning a list of the most similar examples. This approach leverages the principle that similar inputs are likely to produce similar predictions, allowing us to gain insight into the behavior of our model. By examining these similar examples, we can better understand and explain our model's output.

## Feature-based explanation techniques available on Vertex AI

The following table shows different methods of feature-based explanations available in GCP.

| Method | Compatible Vertex AI Model Resources | Example Use Cases |
|---|---|---|
| Sampled Shapley (SHAP) | • Any custom-trained model (running in any prediction container) <br> • AutoML tabular models | • Classification and regression on tabular data |
| Integrated Gradients | • Custom-trained TensorFlow models using TensorFlow prebuilt containers for prediction serving <br> • AutoML image models | • Classification and regression on tabular data <br> • Classification of image data |
| XRAI (eXplanation with Ranked Area Integrals) | • Custom-trained TensorFlow models using TensorFlow prebuilt containers for prediction serving <br> • AutoML image models | • Classification of image data |

Table 8.1 – Feature attribution methods available in GCP. Source: `https://cloud.google.com/vertex-ai/docs/explainable-ai/overview`

Next, we will learn how to use these Vertex AI features to generate explanations for model output.

## Using the model feature importance (SHAP-based) capability with AutoML for tabular data

In the following exercise, we will learn how to use XAI features in Vertex AI to evaluate feature importance in ML models for structured data.

## Exercise 1

Objective: Use Vertex AutoML Tables' **Feature importance** feature to explain global (model-level) and local (sample-level) behavior

Dataset to be used: *Bank Marketing Dataset* (Available in the *Chapter-8* folder within this book's GitHub repository)

Model objective: Predict whether the customer will open a new term deposit or not (**Feature Label – deposit(Yes/No)**)

Follow these steps:

1.  Following the steps shown in *Chapter 5*, create an AutoML classification model to predict the probability of a customer opening the term deposit.

2.  Once the model has been trained, navigate to **Model Registry** | **Your model** | **Your model version** (*1 for new models*) | the **Evaluate** tab.

3.  Scroll down to the **Feature importance** section.

4.  The following **Feature importance** graph shows the relative feature importance of different model features:

## Feature importance

Model feature attribution tells you how much each feature impacted model training. Attribution values are expressed as a percentage; the higher the percentage, the more strongly that feature impacted model training. Model feature attribution is expressed using the Sampled Shapley method. Learn more

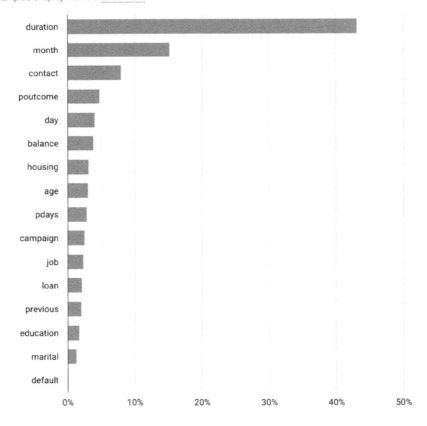

Figure 8.6 – Feature importance graph in Vertex AI AutoML Tables

As shown in the previous graph, the most important features in our model to predict whether a customer will respond to the outreach and open a new term deposit are as follows:

- **Duration**: How long the customer has been with the bank

- **Month**: This could be because of the seasonality of the business

- **Contact** (Method – Cellular/Landline): This could be because of the different communication preferences of different types of customers

- **Outcome**: The outcome of the last promotional outreach to this customer

The following are the least important features as per the preceding graph:

- **Default**: Has the customer defaulted before?

- **Marital (Status)**

- **Education**: Level of education attained

- **Previous**: The number of times contact was made before this campaign

The preceding information can help the data science team better understand the model's behavior and possibly uncover additional insights into their data and customer behavior and provide important guidance for future experiments.

Although performing exploratory data analysis on the training dataset is beyond the scope of this book, for anyone interested, you can look at the notebook (*Chapter 8 – ML Model Explainability – Exercise 1 Addendum*) for the correlation analysis between features and predicted labels.

Here are some insights we can draw from this feature importance information:

- **Duration**: As per correlation analysis in the notebook mentioned previously, duration has the strongest correlation with a predicted label (`Deposit_Signup`), which aligns with the duration being high in terms of feature importance in the preceding graph.

- **Contact**: Similarly, correlation analysis also shows that someone owning a cellular phone has a strong correlation with someone opening a deposit in response to the campaign.

- **Outcome**: Correlation analysis also shows that "Success" has a strong correlation with someone opening a deposit. This means that if someone responded positively to the last campaign, there is a stronger chance of that person being influenced by the current campaign.

- **Months**: In correlation analysis, we can also see that some months (March and May specifically) have a strong correlation with the positive outcome of the campaign. This could relate to AutoML surfacing months as an important feature.

## Exercise 2

Objective: Use Vertex AI's feature attribution features to explain image classification predictions

Dataset to be used: `Fast_Food_Classification_Dataset`

Follow these steps to create the image classification model:

1.  Download and unzip the dataset from Kaggle: `https://www.kaggle.com/datasets/utkarshsaxenadn/fast-food-classification-dataset`

2.  Following the steps shown in *Chapter 5*, create an AutoML image classification dataset using `Fast_Food_Classification_Dataset`. Make sure to select data type and objective as **Image classification (Single-label)**, as shown.

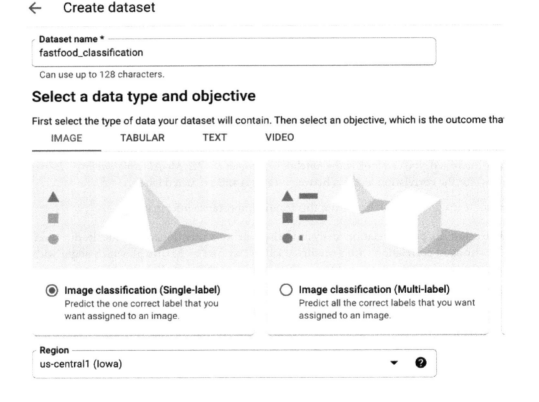

Figure 8.7 – Model objective – Image classification (Single-label)

3.  Once the empty dataset is created, go to the **Browse** tab and add new labels for each food type you plan to include in your model. In this example, we uploaded our favorite fast foods, including burgers, donuts, hot dogs, and pizza, but feel free to use whatever food types you want to use.

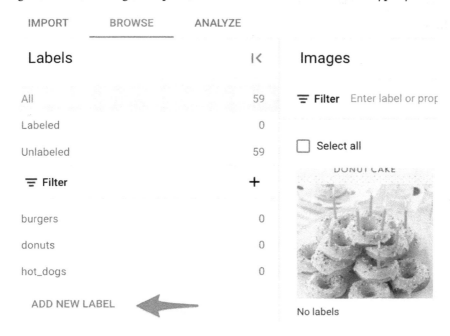

Figure 8.8 – Creating label names

4.  Now let's upload the images of different fast-food types and annotate/label them. You don't need all the available images in the dataset. Just 50 or so images for each food type should suffice.

Repeat the following steps for each food type:

- Upload images – Labeling images one at a time is difficult. To make labeling the images a bit easier, upload images for one food type at a time.

Figure 8.9 – Adding labels to images

As shown, once you have uploaded the images for a particular food item, you can click on **Unlabeled** and then **Select All** to select all the images that need to be labeled. If you upload and label one food type at a time, it ensures that you only select images of one type of food. If you upload all images at once, then clicking on the **Unlabeled** tab would end up selecting ALL unlabeled images, requiring you to manually select images of one type.

5.  Once the images have been selected, click on **ASSIGN LABELS** and select the right food type label. Then click **Save.**

    *Do this process for all different food types.*

6.  Once all images are uploaded and labeled, navigate to the dataset in Vertex AI and go to the **Browse** tab.

7.    Click **TRAIN NEW MODEL**:

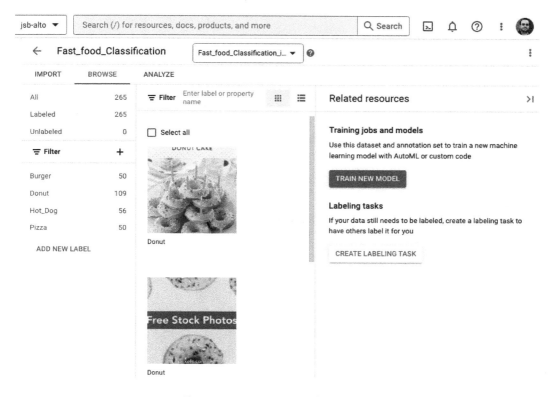

Figure 8.10 – Initiating model training

8.    On the next screen, select the dataset and annotation set you want to use for training the new model. Then, select the following options and click **CONTINUE**:

Objective: **Image classification (Single-label)**

Model training method: **AutoML**

Choose where to use the model: **Cloud**

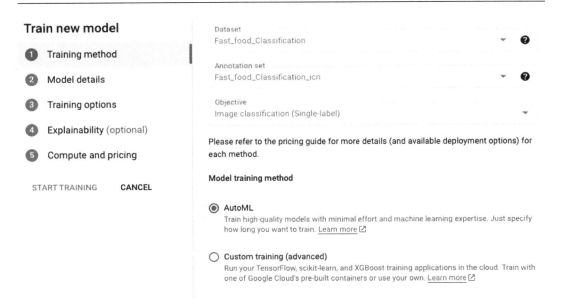

Figure 8.11 – Training configuration/options

9.  On the next screen, select **TRAIN NEW MODEL** and enter the name of the new model. You can leave all other options as is and click **CONTINUE**.

10. On the next screen, select **Default** as the goal and click **CONTINUE**:

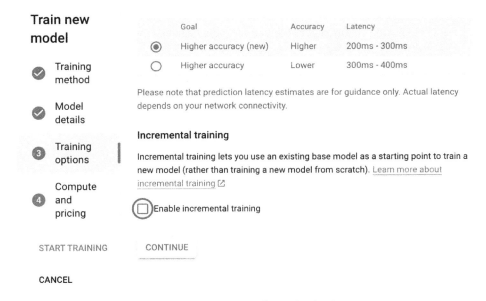

Figure 8.12– Training configuration/options

11. On the next screen (the **Explainability** tab), check the **Generate Explainable Bitmaps** option.

12. On the **Explainability** tab, you can configure key parameters around the two available explainability options shown here (for this exercise, leave the default options as is and click **CONTINUE**):

- **Integrated Gradients**

- **XRAI**

The following are the available configurations:

- **Type**: The type of visualization used, which can be either **OUTLINES** or **PIXELS**. This field is specific to Integrated Gradients and cannot be specified if you are using XRAI. By default, the type is set to **OUTLINES** for Integrated Gradients, which displays regions of attribution. To show per-pixel attribution, set the type to **PIXELS**.

- **Polarity** [*Not available in the UI. Can be used through the API.*]: Polarity indicates the orientation of the spotlighted attributions. By default, it's set to positive, emphasizing regions with the highest positive influences. Essentially, it illuminates pixels that significantly contribute to a positive prediction by the model. Changing the polarity to negative will underscore areas, persuading the model away from predicting the positive class and aiding in pinpointing areas responsible for false negatives. There's also an option to choose both, which provides a comprehensive view by displaying both positive and negative attributions.

- **Upper bound clip percentage**: This setting omits attributions that exceed the defined percentile from the emphasized regions. When combined with other clip parameters, it helps in filtering out distractions, enabling clearer identification of areas with strong attribution.

- **Lower bound clip percentage**: This configuration leaves out attributions falling below the indicated percentile, ensuring only significant regions are highlighted.

- **Color map**: This refers to the color palette chosen for distinguishing highlighted zones. Integrated gradients typically use the pink_green default, where green signifies positive attributions and pink denotes negative ones. On the other hand, XRAI visualizations employ a gradient color scheme, with Viridis as the default. In this setup, the most impactful regions are bathed in yellow, while the less influential areas are shaded in blue. For an exhaustive list of available palettes, consult the **Visualization** message within the API documentation.

- **Overlay type**: This setting defines how the original image is showcased within the visualization. Tweaking the overlay enhances visibility, especially when the inherent properties of the initial image obscure the visualization details.

- **Steps**: The number of steps used to approximate the path integral can be specified here. It is recommended to start with a value of 50 and gradually increase it until the "sum to diff" property falls within the desired error range. The valid range for this value is inclusive between 1 and 100:

 Generate explainable bitmaps for each image in the test set

### Integrated gradients

When you request an explanation using integrated gradients, you'll receive a bitmap with an image overlay showing which pixels contributed to the prediction. Integrated gradients is a pixel-based attribution method that highlights important areas in the image regardless of contrast, making this method ideal for non-natural images such as X-rays.

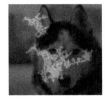

Example

Figure 8.13 – Explainability configurations

- In the **Compute and pricing** tab, set the budget to 8 hours. This specifies the maximum amount of time the training will run.

13. Click **START TRAINING**.

    (A coffee break would be too short, so maybe go prepare a 7-course meal and come back to check on the training status in a few hours!)

    Once the model training is complete, we need to deploy the model to a Vertex AI endpoint by following the next steps.

14. Navigate to **Model Registry** | **Your model** | **Your model version** (*1 for new models*).

15. Navigate to the **Deploy & Test** tab and click **Deploy to endpoint**.

16. Enter the name of the endpoint and click **CONTINUE**:

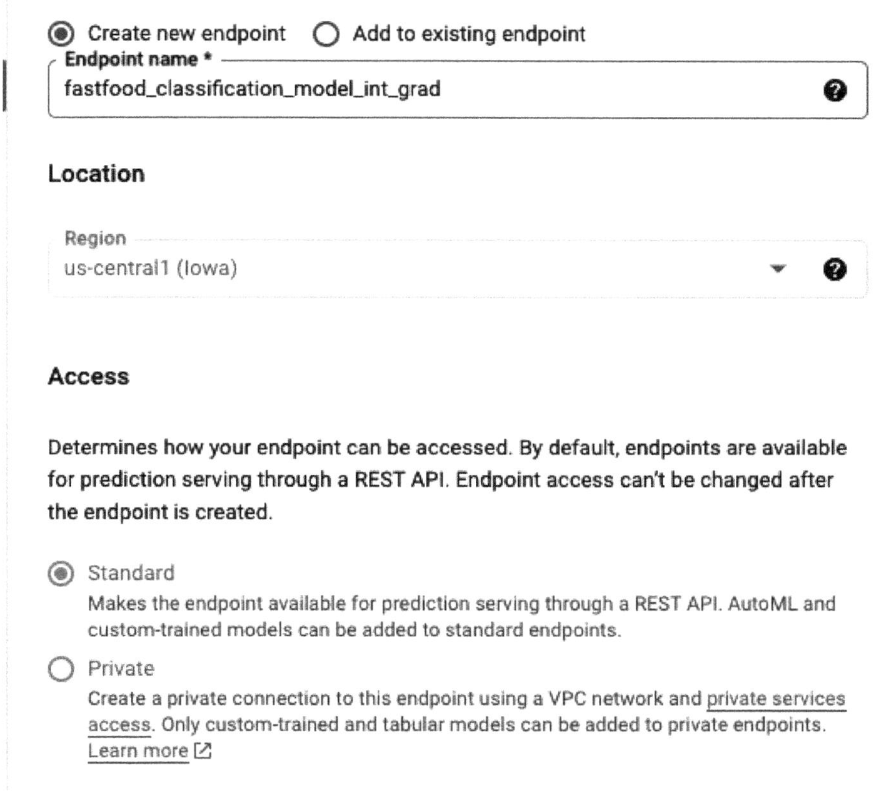

◉ Create new endpoint      ○ Add to existing endpoint

Endpoint name *

fastfood_classification_model_int_grad                                    ❓

**Location**

Region

us-central1 (Iowa)                                                    ▼    ❓

**Access**

Determines how your endpoint can be accessed. By default, endpoints are available for prediction serving through a REST API. Endpoint access can't be changed after the endpoint is created.

◉ Standard

   Makes the endpoint available for prediction serving through a REST API. AutoML and custom-trained models can be added to standard endpoints.

○ Private

   Create a private connection to this endpoint using a VPC network and private services access. Only custom-trained and tabular models can be added to private endpoints. Learn more ↗

Figure 8.14 – Model deployment options

17. In the **Model settings** tab, check **Enable feature attributions for this model** and then click the **EDIT** button underneath to open the **Explainability options** menu:

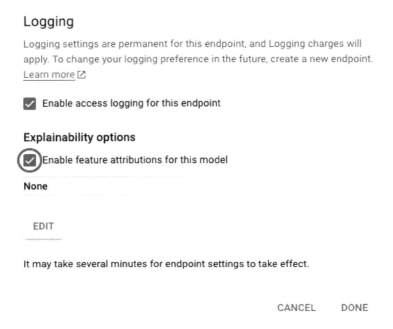

Figure 8.15 – Deployment configuration for explainability

18. In the **Explainability options** menu, select the **Integrated gradients** option since we are first creating an endpoint to test the Integrated Gradients technique. Click **DONE**.

19. Now, repeat these steps to create an endpoint for the **XRAI** explainability option. This time, suffix the name of the endpoint with XRAI, and on the **Explainability options** screen, pick **XRAI**.

20. With that, two endpoints should have been created for the model.

Now, let's test the model by uploading a sample image of donuts and evaluate the prediction and the explanation for the prediction returned by the model:

1. In the model's **Deploy & Test** tab, select the **Integrated Gradients** endpoint by clicking on **Endpoint ID**. Do not click on the endpoint's name as that will take you to the endpoint's settings screen.

2. Click **Upload & Explain** and select an image you want to test.

3. Vertex AI will process the image and present the resulting classification of the image, along with an explanation (the image overlay will show areas of high importance for the image):

Figure 8.16 – Uploaded image

The following screenshot shows the class prediction based on the ML model, along with the Integrated Gradients-based explanation generated by Vertex AI. The explanation image showcases the key areas/pixels in the image that helped the model make the final decision that this is an image of a donut:

Figure 8.17 – Resulting Integrated Gradients explanation and predicted class

4. You can repeat this step with the XRAI endpoint to get explanations using the XRAI technique:

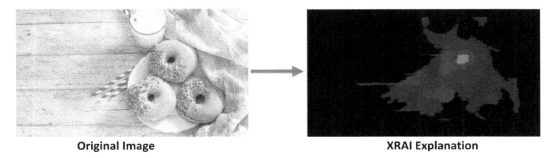

**Original Image**                                    **XRAI Explanation**

Figure 8.18 – XRAI explanation

As you can see, with both the explanation images, which were generated through the Integrated Gradients technique and XRAI technique, the areas/pixels close to the location of donuts in the image are highlighted, and the model seems to be focusing on the correct areas.

Now let's look at example-based explanations where instead of explaining the results based on the features of the input instance, we instead try to explain the results by looking at the examples in the dataset that are similar to the input instance.

## Example-based explanations

Vertex AI's example-based explanation feature uses nearest-neighbor search algorithms to find the closest match to a sample. Essentially, when given an input, Vertex AI identifies and provides a set of examples, typically originating from the training data, that closely resemble the given input. This feature is rooted in the common expectation that inputs with similar attributes will lead to corresponding predictions. Therefore, these identified examples serve as an intuitive way to comprehend and elucidate the workings and decisions of our model.

This method can be extremely helpful in the following scenarios:

- **Identifying mislabeled examples**: If the solution locates data samples or embeddings that are close together in the vector space, but have different labels, then there is a possibility that the data sample is mislabeled.

- **Decision support**: If the predicted labels for new data points are similar to the ground truth labels of other data points appearing close to the new data points in the vector space, then that can help confirm the validity of predictions.

- **Active learning**: In the vector space, you can identify the unlabeled samples appearing close to the labeled samples and add them to the training data with the label of the nearby samples.

The functionality of example-based explanations in Vertex AI can be leveraged by any model offering an embedding – a latent representation – for its inputs. This means that the model should be able to convert the input data into a set of relevant features or vectors in a latent space. This excludes certain types of models, such as tree-based models such as decision trees, from being supported due to their inherent nature of not creating these latent spaces.

## Key steps to use example-based explanations

Here are the key steps:

1.  Enable explanations during model creation: Begin by creating a model with explanations enabled and uploading it to Vertex AI. When you create/import a model, you can set a default configuration for all its explanations using the model's `explanationSpec` field.

    To facilitate the generation of example-based explanations, certain criteria should be met by your model. Two potential scenarios exist for this:

    -   You can implement a **deep neural network** (**DNN**) model, in which case the name of a specific layer or signature should be provided. The output of this layer or signature is then utilized as the latent space.

    -   Alternatively, the model could be designed to directly output embeddings, thus serving as a representation of the latent space.

    This latent space is integral to the process as it houses the example representations that are instrumental in generating explanations.

2.  Deploy the model to an endpoint: Next, create an endpoint resource and deploy your model to it, establishing an accessible channel for interaction.

3.  Explore the generated explanations: Finally, issue explanation requests to the deployed model and scrutinize the provided explanations to understand your model's decision-making processes.

## Exercise 3

Custom train an image classification model to generate real-time predictions and provide example-based explanations – see *Notebook 8.3 – Implementing example-based explanations with Vertex AI* (`https://github.com/PacktPublishing/The-Definitive-Guide-to-Google-Vertex-AI/blob/main/Chapter08/Chapter8_Explainable_AI_example_based.ipynb`)

# Summary

In this chapter, we delved into the world of XAI and its relevance in modern MLOps. We discussed how XAI aids in building trust, ensuring regulatory compliance, debugging and improving models, and addressing ethical considerations.

We explored different explanation techniques for various types of data, including tabular, image, and text data. Techniques such as LIME, SHAP, permutation feature importance, and others were discussed for tabular data. For image data, methods such as Integrated Gradients and XRAI were explained, while text-specific LIME was presented for text data.

This chapter also provided an overview of the XAI features available in GCP, including both feature-based and example-based explanations.

At this point, you should have gained a good understanding of XAI, its importance, various techniques, and practical applications in the context of Vertex AI. As the field of AI continues to evolve, the role of XAI in creating transparent, trustworthy, and fair ML models will only grow. As MLOps practitioners, having these skills will be crucial in leading ethical and responsible AI adoption.

In the next chapter, we will go over various Vertex AI tools that can help you iterate through model hyperparameters to improve the performance of your ML solutions.

# References

https://cloud.google.com/vertex-ai/docs/explainable-ai/overview

Munn, Michael; Pitman, David. *Explainable AI for Practitioners*. O'Reilly Media.

# 9

# Model Optimizations – Hyperparameter Tuning and NAS

We have now become quite familiar with some of the Vertex AI offerings related to managing data, training no-code and low-code models, and launching large-scale custom model training jobs (with metadata tracking and monitoring capabilities). As ML practitioners, we know that it is highly unlikely that the first model we train would be the best model for a given use case and dataset. Thus, in order to find the best model (which is the most accurate and least biased), we often use different model optimization techniques. **Hyperparameter Tuning (HPT)** and **Neural Architecture Search (NAS)** are two such model optimization techniques. In this chapter, we will learn how to configure and launch model optimization experiments using Vertex AI on Google Cloud.

In this chapter, we will first learn about the importance of model optimization techniques such as HPT and then learn how to quickly set up and launch HPT jobs within Google Vertex AI. We will also understand how NAS works and how it is different from HPT. The topics covered in this chapter are as follows:

- What is HPT and why is it important?

- Setting up HPT jobs on Vertex AI

- What is NAS and how is it different from HPT?

- NAS on Vertex AI overview

## Technical requirements

The code examples shown in this chapter can be found in the following GitHub repo: `https://github.com/PacktPublishing/The-Definitive-Guide-to-Google-Vertex-AI/tree/main/Chapter09`

# What is HPT and why is it important?

Hyperparameter tuning, or HPT for short, is a popular model optimization technique that is very commonly used across ML projects. In this section, we will learn about hyperparameters, the importance of tuning them, and different methods of finding the best hyperparameters for a machine learning algorithm.

## What are hyperparameters?

When we train an ML system, we basically have three kinds of data – input data, model parameters, and model hyperparameters. Input data refers to our training or test data that is associated with the problem we are solving. Model parameters are the variables that we modify during the model training process and we try to adjust them to fit the training data. Model hyperparameters, on the other hand, are variables that govern the training process itself. These hyperparameters are fixed before we start to train our model. For example, learning rate, optimizer, batch size, number of hidden layers in a neural network, and the max depth in a tree-based algorithm are some examples of model hyperparameters.

## Why HPT?

How well your machine learning model will perform largely depends upon the hyperparameters you choose before training. The values of hyperparameters can make all the difference to performance metrics (such as accuracy), training time, bias, fairness, and so on for your model. Hyperparameter tuning or HPT is a model optimization technique that chooses a set of optimal hyperparameters for a learning algorithm. The same ML algorithm can require totally different values of hyperparameters to generalize for different data patterns. There is an objective function associated with every HPT job that it tries to optimize (minimize or maximize) and it returns the values of hyperparameters that achieve that optimal value. This objective function can be similar to the model training objective (e.g., loss function) or it can be a completely new metric.

We run model optimization operations such as HPT or NAS when we are at a stage where our final model (i.e., XGBoost) is fixed and we have a fixed test set for which we want to optimize the hyperparameters of our chosen model. A typical HPT job runs multiple trials with different sets of hyperparameters and returns the hyperparameters that lead to the best trial. The best trial here represents the trial that optimizes the objective function associated with the HPT job.

## Search algorithms

While running HPT, we have to decide what kind of search algorithm we want to run over our hyperparameter space. There are multiple different kinds of search algorithms that we can choose from based on our needs. A few commonly used approaches are as follows:

- Grid search

- Random search

- Bayesian optimization

Let's discuss these approaches!

### Grid search

The traditional way of performing HPT has been grid search, which is basically an exhaustive search over a manually specified search space. A grid search must be provided with a performance metric that it tries to calculate over all possible sets of hyperparameter combinations, measured over a hold-out validation set (or cross-validation on a training set). As it runs all possible combinations of provided hyperparameter ranges, it is important to set those ranges carefully and with discrete values. As grid search runs all the trials independently, it can be parallelized for faster outcomes.

### Random search

Instead of trying all combinations sequentially and exhaustively like grid search, random search selects hyperparameters randomly from the provided search space during each trial. As it chooses hyperparameter values randomly, it also generalizes to continuous spaces along with discrete spaces, as discussed above. Random search again is highly parallelizable as all the trials are independent. Despite its simplicity, random search is one of the most important baselines to test new optimization or search techniques against.

### Bayesian optimization

Unlike grid search and random search, the Bayesian optimization method builds a probabilistic model of the function that maps hyperparameter values to the HPT objective function. Thus, at each new trial, it learns better about the direction it should take to find the optimal hyperparameters for the given objective function on a fixed validation set. It tries to balance exploration and exploitation and has been shown to obtain better results than the preceding techniques in fewer trials. But as it learns from the ongoing trials, it often runs the trials iteratively (thus it's less parallelizable).

Now that we have a good understanding of HPT, let's understand how to set up and launch HPT jobs on Vertex AI.

## Setting up HPT jobs on Vertex AI

In this section, we will learn how to set up HPT jobs with Vertex AI. We will use the same neural network model experiment from *Chapter 7, Training Fully Custom ML Models with Vertex AI*, and optimize its hyperparameters to get the best model settings.

The first step is to create a new Jupyter Notebook in Vertex AI Workbench and import useful libraries:

```
import numpy as np
import glob
import matplotlib.pyplot as plt
import os
import google.cloud.aiplatform as aiplatform
```

```
from google.cloud.aiplatform import hyperparameter_tuning as hpt
from datetime import datetime
TIMESTAMP = datetime.now().strftime("%Y%m%d%H%M%S")
%matplotlib inline
```

Next, we set up project configurations:

```
PROJECT_ID='************'
REGION='us-west2'
SERVICE_ACCOUNT='417xxxxxxxxx7-compute@developer.gserviceaccount.com'
BUCKET_URI='gs://my-training-artifacts'
```

Then, we initialize the Vertex AI SDK:

```
aiplatform.init(project=PROJECT_ID, location=REGION, \
    staging_bucket=BUCKET_URI)
```

The next step is to containerize the full training application code. We will put our full training code into a Python file, `task.py`, here. The `task.py` file should have an entire flow, including the following:

- Loading and preparing the training data

- Defining the model architecture

- Training the model (running a trial with given hyperparameters as args)

- Saving the model (optional)

- Passing the training trial output to the `hypertune()` method

The training script should have a list of hyperparameters that it wants to tune, defined as arguments:

```
def get_args():
    '''Parses args. Must include all hyperparameters you want to
tune.'''

    parser = argparse.ArgumentParser()
    parser.add_argument(
      '--epochs',
      required=True,
      type=int,
      help='training epochs')
    parser.add_argument(
      '--steps_per_epoch',
      required=True,
      type=int,
      help='steps_per_epoch')
```

Similarly, we have other important hyperparameters, such as learning rate, batch size, loss function, and so on:

```
parser.add_argument(
  '--learning_rate',
  required=True,
  type=float,
  help='learning rate')
parser.add_argument(
  '--batch_size',
  required=True,
  type=int,
  help='training batch size')
parser.add_argument(
  '--loss',
  required=True,
  type=str,
  help='loss function')

args = parser.parse_args()
return args
```

The script should have a function that loads and prepares the training and validation datasets:

```
def make_datasets_unbatched():
    # Load train, validation and test sets
    dest = 'gs://data-bucket-417812395597/'
    train_x = np.load(BytesIO(
        file_io.read_file_to_string(dest+'train_x', \
            binary_mode=True)
    ))
    train_y = np.load(BytesIO(
        file_io.read_file_to_string(dest+'train_y', \
            binary_mode=True)
    ))
```

Similarly, the validation and test data parts are loaded:

```
    val_x = np.load(BytesIO(
        file_io.read_file_to_string(dest+'val_x', \
            binary_mode=True)
    ))
    val_y = np.load(BytesIO(
        file_io.read_file_to_string(dest+'val_y', \
            binary_mode=True)
```

```
    ))
    test_x = np.load(BytesIO(
        file_io.read_file_to_string(dest+'test_x', \
            binary_mode=True)
    ))
    test_y = np.load(BytesIO(
        file_io.read_file_to_string(dest+'test_y', \
            binary_mode=True)
    ))
    return train_x, train_y, val_x, val_y, test_x, test_y
```

The preceding function loads the already prepared dataset from GCS. We can refer to *Chapter 7, Training Fully Custom ML Models with Vertex AI*, to fully understand the data preparation part.

Next, we define the **TensorFlow** (**TF**) model architecture:

```
def tf_model():
    black_n_white_input = tensorflow.keras.layers.Input(shape=(80, 80,
1))

    enc = black_n_white_input
```

We then define the encoder part of the model:

```
    #Encoder part
    enc = tensorflow.keras.layers.Conv2D(
        32, kernel_size=3, strides=2, padding='same'
    )(enc)
    enc = tensorflow.keras.layers.LeakyReLU(alpha=0.2)(enc)
    enc = tensorflow.keras.layers.BatchNormalization(momentum=0.8)
(enc)

    enc = tensorflow.keras.layers.Conv2D(
        64, kernel_size=3, strides=2, padding='same'
    )(enc)
    enc = tensorflow.keras.layers.LeakyReLU(alpha=0.2)(enc)
    enc = tensorflow.keras.layers.BatchNormalization(momentum=0.8)
(enc)
```

Similarly, we will define two more encoder layers with an increasing number of filters, a kernel size of 3, and a stride of 2 so that we can compress the image into important features:

```
    enc = tensorflow.keras.layers.Conv2D(
        128, kernel_size=3, strides=2, padding='same'
    )(enc)
    enc = tensorflow.keras.layers.LeakyReLU(alpha=0.2)(enc)
```

```
    enc = tensorflow.keras.layers.BatchNormalization(momentum=0.8)
(enc)

    enc = tensorflow.keras.layers.Conv2D(
        256, kernel_size=1, strides=2, padding='same'
    )(enc)
    enc = tensorflow.keras.layers.LeakyReLU(alpha=0.2)(enc)
    enc = tensorflow.keras.layers.Dropout(0.5)(enc)
```

Define the decoder part of the TF model within the same function:

```
    #Decoder part
    dec = enc

    dec = tensorflow.keras.layers.Conv2DTranspose(
        256, kernel_size=3, strides=2, padding='same'
    )(dec)
    dec = tensorflow.keras.layers.Activation('relu')(dec)
    dec = tensorflow.keras.layers.BatchNormalization(momentum=0.8)
(dec)

    dec = tensorflow.keras.layers.Conv2DTranspose(
        128, kernel_size=3, strides=2, padding='same'
    )(dec)
    dec = tensorflow.keras.layers.Activation('relu')(dec)
    dec = tensorflow.keras.layers.BatchNormalization(momentum=0.8)
(dec)
```

As we can see, the decoder design is almost opposite to the encoder part. Here, we re-create the image from compressed features by using multiple layers of transpose convolutions and reducing the channels gradually to 3 to generate the final color image output:

```
    dec = tensorflow.keras.layers.Conv2DTranspose(
        64, kernel_size=3, strides=2, padding='same'
    )(dec)
    dec = tensorflow.keras.layers.Activation('relu')(dec)
    dec = tensorflow.keras.layers.BatchNormalization(momentum=0.8)
(dec)

    dec = tensorflow.keras.layers.Conv2DTranspose(
        32, kernel_size=3, strides=2, padding='same'
    )(dec)
    dec = tensorflow.keras.layers.Activation('relu')(dec)
    dec = tensorflow.keras.layers.BatchNormalization(momentum=0.8)
(dec)

    dec = tensorflow.keras.layers.Conv2D(
```

```
        3, kernel_size=3, padding='same'
) (dec)
```

Add a `tanh` activation function to get the final colored output image:

```
color_image = tensorflow.keras.layers.Activation('tanh')(dec)

return black_n_white_input, color_image
```

Also, add a function to build and compile the TF model:

```
# Build the and compile TF model
def build_and_compile_tf_model(loss_fn, learning_rate):
    black_n_white_input, color_image = tf_model()
    model = tensorflow.keras.models.Model(
        inputs=black_n_white_input,
        outputs=color_image
    )
    _optimizer = tensorflow.keras.optimizers.Adam(
        learning_rate=learning_rate,
        beta_1=0.5
    )
    model.compile(
        loss=loss_fn,
        optimizer=_optimizer
    )
    return model
```

Finally, add a `main` function that trains the model and provides the hyperparameter tuning metric value to the `hypertune()` function. In our case, we will be optimizing the loss over the validation dataset. See the following snippet:

```
def main():
    args = get_args()
```

Set up the configurations and load the data:

```
    NUM_WORKERS = strategy.num_replicas_in_sync
    # Global batch size should be scaled as per the number     # of
workers used in training.    GLOBAL_BATCH_SIZE = args.batch_size *
NUM_WORKERS
    MODEL_DIR = os.getenv("AIP_MODEL_DIR")

    train_x, train_y, val_x, val_y, _, _ = \
        make_datasets_unbatched()
```

Now, let's build the TF model and fit it on the training data:

```
with strategy.scope():
    # Creation of dataset, and model building/compiling need to be
within
    # `strategy.scope()`.
    model = build_and_compile_tf_model(args.loss, \
        args.learning_rate)

history = model.fit(
    train_x,
    train_y,
    batch_size=GLOBAL_BATCH_SIZE,
    epochs=args.epochs,
    steps_per_epoch=args.steps_per_epoch,
    validation_data=(val_x, val_y),
)
model.save(MODEL_DIR)
```

Use `hypertune` to define and report hyperparameter tuning metrics to the HPT algorithm:

```
# DEFINE HPT METRIC
hp_metric = history.history['val_loss'][-1]

hpt = hypertune.HyperTune()
hpt.report_hyperparameter_tuning_metric(
    hyperparameter_metric_tag='val_loss',
    metric_value=hp_metric,
    global_step=args.epochs)
```

If we put this all into a single Python file, our `task.py` file should look something like the following:

```
%%writefile task.py
# Single, Mirror and Multi-Machine Distributed Training
```

Load all the dependencies for our task:

```
import tensorflow as tf
import tensorflow
from tensorflow.python.client import device_lib
import argparse
import os
import sys
from io import BytesIO
import numpy as np
```

```
from tensorflow.python.lib.io import file_io
import hypertune
```

Parse arguments where we define the hyperparameters used for tuning:

```
def get_args():
    '''Parses args. Must include all hyperparameters you want to
tune.'''

    parser = argparse.ArgumentParser()
    parser.add_argument(
      '--epochs',
      required=True,
      type=int,
      help='training epochs')
    parser.add_argument(
      '--steps_per_epoch',
      required=True,
      type=int,
      help='steps_per_epoch')
```

Define a few more hyperparameter-related arguments for tuning the learning rate, batch size, and loss functions:

```
    parser.add_argument(
      '--learning_rate',
      required=True,
      type=float,
      help='learning rate')
    parser.add_argument(
      '--batch_size',
      required=True,
      type=int,
      help='training batch size')
    parser.add_argument(
      '--loss',
      required=True,
      type=str,
      help='loss function')

    args = parser.parse_args()
    return args
```

Set up configurations for training purposes:

```
print('Python Version = {}'.format(sys.version))
print('TensorFlow Version = {}'.format(tf.__version__))
print('TF_CONFIG = {}'.format(os.environ.get('TF_CONFIG', \
    'Not found')))
print('DEVICES', device_lib.list_local_devices())
```

Define configuration settings for training distribution strategies based on the requirements – it can be a single, mirror, or multi-worker strategy:

```
DISTRIBUTE='single'
if DISTRIBUTE == 'single':
    if tf.test.is_gpu_available():
        strategy = tf.distribute.OneDeviceStrategy(device="/gpu:0")
    else:
        strategy = tf.distribute.OneDeviceStrategy(device="/cpu:0")
# Single Machine, multiple compute device
elif DISTRIBUTE == 'mirror':
    strategy = tf.distribute.MirroredStrategy()
# Multiple Machine, multiple compute device
elif DISTRIBUTE == 'multi':
    strategy = tf.distribute.experimental.
MultiWorkerMirroredStrategy()print('num_replicas_in_sync = {}'.
format(strategy.num_replicas_in_sync))
```

Load and prepare the training, validation, and test partitions of data from the GCS bucket:

```
# Preparing dataset
BUFFER_SIZE = 10000

def make_datasets_unbatched():
    # Load train, validation and test sets
    dest = 'gs://data-bucket-417812395597/'
    train_x = np.load(BytesIO(
        file_io.read_file_to_string(dest+'train_x', \
            binary_mode=True)
    ))
    train_y = np.load(BytesIO(
        file_io.read_file_to_string(dest+'train_y', \
            binary_mode=True)
    ))
```

Similarly, load the validation and test partitions:

```
val_x = np.load(BytesIO(
    file_io.read_file_to_string(dest+'val_x', \
        binary_mode=True)
))
val_y = np.load(BytesIO(
    file_io.read_file_to_string(dest+'val_y', \
        binary_mode=True)
))
test_x = np.load(BytesIO(
    file_io.read_file_to_string(dest+'test_x', \
        binary_mode=True)
))
test_y = np.load(BytesIO(
    file_io.read_file_to_string(dest+'test_y', \
        binary_mode=True)
))
return train_x, train_y, val_x, val_y, test_x, test_y
```

Define the TF model architecture for converting a black-and-white image to a color image:

```
def tf_model():
    black_n_white_input = tensorflow.keras.layers.Input(shape=(80, 80,
1))

    enc = black_n_white_input
```

Define the encoder part of the model:

```
    #Encoder part
    enc = tensorflow.keras.layers.Conv2D(
        32, kernel_size=3, strides=2, padding='same'
    )(enc)
    enc = tensorflow.keras.layers.LeakyReLU(alpha=0.2)(enc)
    enc = tensorflow.keras.layers.BatchNormalization(momentum=0.8)
(enc)

    enc = tensorflow.keras.layers.Conv2D(
        64, kernel_size=3, strides=2, padding='same'
    )(enc)
    enc = tensorflow.keras.layers.LeakyReLU(alpha=0.2)(enc)
    enc = tensorflow.keras.layers.BatchNormalization(momentum=0.8)
(enc)
```

Similarly, we will define two more encoder layers with an increasing number of filters, a kernel size of 3, and a stride of 2 so that we can compress the image into important features:

```
enc = tensorflow.keras.layers.Conv2D(
    128, kernel_size=3, strides=2, padding='same'
)(enc)
enc = tensorflow.keras.layers.LeakyReLU(alpha=0.2)(enc)
enc = tensorflow.keras.layers.BatchNormalization(momentum=0.8)
(enc)

enc = tensorflow.keras.layers.Conv2D(
    256, kernel_size=1, strides=2, padding='same'
)(enc)
enc = tensorflow.keras.layers.LeakyReLU(alpha=0.2)(enc)
enc = tensorflow.keras.layers.Dropout(0.5)(enc)
```

Define the decoder part of the model:

```
 #Decoder part
dec = enc
dec = tensorflow.keras.layers.Conv2DTranspose(
    256, kernel_size=3, strides=2, padding='same'
)(dec)
dec = tensorflow.keras.layers.Activation('relu')(dec)
dec = tensorflow.keras.layers.BatchNormalization(momentum=0.8)
(dec)

dec = tensorflow.keras.layers.Conv2DTranspose(
    128, kernel_size=3, strides=2, padding='same'
)(dec)
dec = tensorflow.keras.layers.Activation('relu')(dec)
dec = tensorflow.keras.layers.BatchNormalization(momentum=0.8)
(dec)
```

As we can see, the decoder design is almost opposite to the encoder part. Here, we re-create the image from compressed features by using multiple layers of transpose convolutions and reduce the channels gradually to 3 to generate the final color image output:

```
dec = tensorflow.keras.layers.Conv2DTranspose(
    64, kernel_size=3, strides=2, padding='same'
)(dec)
dec = tensorflow.keras.layers.Activation('relu')(dec)
dec = tensorflow.keras.layers.BatchNormalization(momentum=0.8)
(dec)

dec = tensorflow.keras.layers.Conv2DTranspose(
```

```
        32, kernel_size=3, strides=2, padding='same'
    )(dec)
    dec = tensorflow.keras.layers.Activation('relu')(dec)
    dec = tensorflow.keras.layers.BatchNormalization(momentum=0.8)
(dec)
    dec = tensorflow.keras.layers.Conv2D(
        3, kernel_size=3, padding='same'
    )(dec)
```

Finally, generate the color image output by using the `tanh` activation function:

```
    color_image = tensorflow.keras.layers.Activation('tanh')(dec)

    return black_n_white_input, color_image
```

The following function will build and compile the TF model for us:

```
`# Build the and compile TF model
def build_and_compile_tf_model(loss_fn, learning_rate):
    black_n_white_input, color_image = tf_model()
    model = tensorflow.keras.models.Model(
        inputs=black_n_white_input,
        outputs=color_image
    )
    _optimizer = tensorflow.keras.optimizers.Adam(
        learning_rate=learning_rate,
        beta_1=0.5
    )
    model.compile(
        loss=loss_fn,
        optimizer=_optimizer
    )
    return model
```

Now, let's define the main function to start executing our training and tuning task. Here, the num_ replicas_in_sync parameter defines how many training tasks are running in parallel on different workers in a multi-worker training strategy:

```
def main():
    args = get_args()

    NUM_WORKERS = strategy.num_replicas_in_sync
    # Here the batch size scales up by number of workers since
    # `tf.data.Dataset.batch` expects the global batch size.
    GLOBAL_BATCH_SIZE = args.batch_size * NUM_WORKERS
    MODEL_DIR = os.getenv("AIP_MODEL_DIR")
```

Load the training and validation data to start training our TF model:

```
train_x, train_y, val_x, val_y, _, _ = \
    make_datasets_unbatched()

with strategy.scope():
    # Creation of dataset, and model building/compiling need to be
within
    # `strategy.scope()`.
    model = build_and_compile_tf_model(args.loss, \
        args.learning_rate)

history = model.fit(
    train_x,
    train_y,
    batch_size=GLOBAL_BATCH_SIZE,
    epochs=args.epochs,
    steps_per_epoch=args.steps_per_epoch,
    validation_data=(val_x, val_y),
)
model.save(MODEL_DIR)
```

Finally, define the HPT metric with the help of the `hypertune` package:

```
# DEFINE HPT METRIC
hp_metric = history.history['val_loss'][-1]

hpt = hypertune.HyperTune()
hpt.report_hyperparameter_tuning_metric(
    hyperparameter_metric_tag='val_loss',
    metric_value=hp_metric,
    global_step=args.epochs)


if __name__ == "__main__":
    main()
```

Next, we create a staging bucket in GCS that will be used for storing artifacts such as trial outcomes from our HPT job:

```
BUCKET_URI = "gs://hpt-staging"  # @param {type:"string"}

if BUCKET_URI == "" or BUCKET_URI is None or BUCKET_URI == "gs://
[your-bucket-name]":
```

```
    BUCKET_URI = "gs://" + PROJECT_ID + "aip-" + TIMESTAMP

! gsutil mb -l {REGION} -p {PROJECT_ID} {BUCKET_URI}

GCS_OUTPUT_BUCKET = BUCKET_URI + "/output/"
```

The next step is to containerize the entire training code defined in the task.py file. The hyperparameter tuning job will use this container to launch different trials with different hyperparameters as arguments:

```
%%writefile Dockerfile
FROM gcr.io/deeplearning-platform-release/tf2-gpu.2-8
WORKDIR /
# Installs hypertune library
RUN pip install cloudml-hypertune

# Copies the trainer code to the Docker image.
COPY task.py .
# Sets up the entry point to invoke the trainer.
ENTRYPOINT ["python", "-m", "task"]
```

Our Dockerfile is ready – let's build and push the Docker image to **Google Container Registry (GCR)**:

```
PROJECT_NAME="*******-project"
IMAGE_URI = (
    f"gcr.io/{PROJECT_NAME}/example-tf-hptune:latest"
)
! docker build ./ -t $IMAGE_URI
! docker push $IMAGE_URI
```

Now we have a container image ready with all the training code that we need. Let's configure the HPT job.

First, we define the type of machine we want our trials to run on. The machine specification will depend upon the size of the model and training dataset. As this is a small experiment, we will use the n1-standard-8 machine to run it:

```
# The spec of the worker pools including machine type and Docker image
# Be sure to replace PROJECT_ID in the `image_uri` with your project.

worker_pool_specs = [
    {
        "machine_spec": {
            "machine_type": "n1-standard-8",
            "accelerator_type": None,
            "accelerator_count": 0,
        },
```

```
        "replica_count": 1,
        "container_spec": {
            "image_uri": f"gcr.io/{PROJECT_NAME}/example-tf-
hptune:latest"
        },
    }
]
```

Note that, within the worker pool spec, we have also passed the path to the training image that we created.

Next, we will define the parameter space that our job will use to find the best hyperparameters:

```
# Dictionary representing parameters to optimize.
# The dictionary key is the parameter_id, which is passed into your
training
# job as a command line argument,
# And the dictionary value is the parameter specification of the
metric.
parameter_spec = {
    "learning_rate": hpt.DoubleParameterSpec(min=0.0001, \
        max=0.001, scale="log"),
    "epochs": hpt.DiscreteParameterSpec(values=[10, 20, \
        30], scale=None),
    "steps_per_epoch": hpt.IntegerParameterSpec(min=100, \
        max=300, scale="linear"),
    "batch_size": hpt.DiscreteParameterSpec(values=[16,32,\
        64], scale=None),
    "loss": hpt.CategoricalParameterSpec(["mse"]), # we can add other
loss values
}
```

The parameter space should be carefully defined based on best practices and prior knowledge so that the HPT job doesn't have to perform unnecessary trials over unimportant hyperparameter ranges.

Next, we need to define the metric specifications. In our case, as we are trying to optimize the validation loss value, we would like to minimize it. In the case of accuracy, we should have maximized our metric:

```
metric_spec = {"val_loss": "minimize"}
```

Vertex AI HPT jobs use the Bayesian optimization approach by default to find the best hyperparameters for our settings. We also have the option to use other optimization approaches. As Bayesian optimization works best for most cases, we will be using it in our experiment.

Next, we define the custom job that will run our hyperparameter tuning trials:

```
my_custom_job = aiplatform.CustomJob(
    display_name="example-tf-hpt-job",
    worker_pool_specs=worker_pool_specs,
    staging_bucket=GCS_OUTPUT_BUCKET,
)
```

Finally, we define the HPT job that will launch the trials using the preceding custom job:

```
hp_job = aiplatform.HyperparameterTuningJob(
    display_name="example-tf-hpt-job",
    custom_job=my_custom_job,
    metric_spec=metric_spec,
    parameter_spec=parameter_spec,
    max_trial_count=5,
    parallel_trial_count=3,
)
```

Note that the `max_trial_count` and `parallel_trial_count` parameters are important here:

- `max_trial_count`: You need to put an upper bound on the number of trials the service will run. More trials generally lead to better results, but there will be a point of diminishing returns, after which additional trials have little or no effect on the metric you're trying to optimize. It is best practice to start with a smaller number of trials and get a sense of how impactful your chosen hyperparameters are before scaling up.

- `parallel_trial_count`: If you use parallel trials, the service provisions multiple training processing clusters. Increasing the number of parallel trials reduces the amount of time the hyperparameter tuning job takes to run; however, it can reduce the effectiveness of the job overall. This is because the default tuning strategy uses the results of previous trials to inform the assignment of values in subsequent trials. If we keep the parallel trial count equal to the number of maximum trials, then all trials will start running in parallel, and we will end up running a "random parameter search" here as there will not be any scope of learning from the performance of previous trials.

Now that we are all set, we can launch the HPT job:

```
hp_job.run()
```

As soon as we launch the job, it provides us with a link to the Cloud console UI, where we can monitor the progress of our HPT trials and jobs. The Cloud console UI looks something similar to what's shown in *Figure 9.1*.

Figure 9.1 – HPT job monitoring within the Cloud console UI

Now that we have successfully understood and launched an HPT job on Vertex AI, let's jump to the next section and understand the NAS model optimization technique.

## What is NAS and how is it different from HPT?

**Artificial Neural Networks** or **ANNs** are widely used today for solving complex ML problems. Most of the time, these network architectures are hand-designed by ML experts, which may not be optimal every time. **Neural Architecture Search** or **NAS** is a technique that automates the process of designing neural network architectures that usually outperform hand-designed networks.

Although both HPT and NAS are used as model optimization techniques, there are certain differences in how they both work. HPT assumes a given architecture and focuses on optimizing the hyperparameters that lead to the best model. HPT optimizes hyperparameters such as learning rate, optimizer, batch size, activation function, and so on. NAS, on the other hand, focuses on optimizing architecture-specific parameters (in a way, it automates the process of designing a neural network architecture). NAS optimizes parameters such as the number of layers, number of units, types of connections between layers, and so on. Using NAS, we can search for optimal neural architectures in terms of accuracy, latency, memory, a combination of these, or a custom metric.

NAS usually works with a relatively larger search space than HPT and controls different aspects of network architectures. However, the underlying problem addressed is the same as HPT optimization. There are many NAS-based optimization approaches, but on a high level, any NAS approach has three main components, as follows:

- Search space
- Optimization method
- Evaluation method

Let's learn more about each of these components.

## Search space

This component controls the set of possible neural architectures to consider. The search space is often problem-specific, as a vision-related problem would have the possibility of having **Convolutional Neural Network** (**CNN**) layers as well. However, the process of identifying the best architecture is automated by NAS. Carefully designing these search spaces still depends upon human expertise.

## Optimization method

This component decides how to navigate the search space to find the best possible architecture for a given application. Many different optimization methods have been applied to NAS, such as **reinforcement learning** (**RL**), Bayesian optimization, gradient-based optimization, evolutionary search, and so on. Each of these methods has its own way of evaluating the architectures, but the high-level goal is to focus on the area of the search space that provides better performance. This aspect of NAS is quite similar to HPT optimization methods.

## Evaluation method

The evaluation method is a component that is used for assessing the quality of architectures designed by the chosen optimization method. One simple way to evaluate neural architecture is to fully train it, but this method is quite computationally expensive. Alternatively, to make NAS more efficient, partial training and evaluation methods have been developed. In order to provide cheaper heuristic measures of neural network quality, some evaluation methods have been developed. These evaluation methods are quite specific to NAS and exploit the basic structure of a neural network to estimate the quality of a network. Some examples of these methods include weight-sharing, hypernetworks, network morphism, and so on. These NAS-specific evaluation methods are practically way cheaper than full training.

We now have a good understanding of the NAS optimization method and how it works. Next, let's explore the Vertex AI offering and its features for launching NAS on Google Cloud.

# NAS on Vertex AI overview

Vertex AI NAS is an optimization technique that can be leveraged to find the best neural network architecture for a given ML use case. NAS-based optimization searches for the best network in terms of accuracy but can also be augmented with other constraints such as latency, memory, or a custom metric as per the requirements. In general, the search space of possible neural networks can be quite large and NAS may support a search space as large as $10^{20}$. In the past few years, NAS has been able to successfully generate some state-of-the-art computer vision network architectures, including NASNet, MNasNet, EfficientNet, SpineNet, NAS-FPN, and so on.

It may seem complex, but NAS features are quite flexible and easy to use. A beginner can leverage prebuilt modules for search spaces, trainer scripts, and Jupyter notebooks to start exploring Vertex AI NAS on a custom dataset. If you are an expert, you could potentially develop custom trainer scripts, custom search spaces, custom evaluation methods, and even develop applications for non-vision-based use cases.

Vertex AI can be leveraged to explore the full set of NAS features for our customized architectures and use cases. Here is what Vertex AI provides us with to help in implementing NAS more conveniently:

- Vertex AI provides a NAS-specific language that can be leveraged to define a custom search space to try out the desired set of possible neural network architectures and integrate this space with our custom trainer scripts.

- Pre-built state-of-the-art search spaces and a trainer that are ready to use and can run on a GPU.

- A pre-defined NAS controller that samples our custom-defined search space to find the best neural network architecture.

- A set of prebuilt libraries and Docker images that can be leveraged to calculate latency, FLOPS (Floating-point operations per second), or memory usage on a custom hardware setting.

- Google Cloud provides tutorials to explain the usage of NAS. It also provides examples and guidance for setting up NAS for PyTorch-based applications efficiently.

- Pre-built tools to design proxy tasks.

- There is library support that can be leveraged to report custom-defined metrics and perform analysis on them.

- The Google Cloud console is very helpful in monitoring and managing NAS jobs. We also get some easy-to-use example notebooks to kick-start the search.

- Management of CPU/GPU resource usage on the basis of per project or per job, with the help of a prebuilt library.

- A NAS client to build Docker images, launch NAS jobs, and resume an old NAS search job that is Python-based.

- Customer support is Google Cloud console UI-based.

These features can help us in setting up a custom NAS job without putting in too much effort. Now let's discuss some of the best practices while working with NAS.

## NAS best practices

The important thing to note here is that NAS is not an optimization method that we should apply to all our ML problems. There are certain things to keep in mind before deciding to run a NAS job for our use case. Some of these best practices are as follows:

- NAS is not meant for tuning the hyperparameters of a model. It only performs an architecture search and it is not advised to compare the results of these two methods. In some setups, HPT can be followed by NAS.

- NAS is not recommended for smaller or highly imbalanced datasets.

- NAS is expensive, so unless we can spend a few thousand dollars without extremely high expectations, it's not meant for us.

- You should first try other traditional and conventional machine learning methods and techniques such as hyperparameter tuning. You should use neural architecture search only if you don't see further gains with traditional methods.

Setting up NAS jobs on Vertex AI is not very complex, thanks to the prebuilt assets and publicly released code examples. With these prebuilt features, examples, and best practices, we should be able to set up a custom NAS job that can help us find an optimal architecture to meet our project goals.

# Summary

In this chapter, we discussed the importance of applying model optimization techniques to get the best performance for our application. We learned about two model optimization methods – HPT and NAS, with their similarities and differences. We also learned how to set up and launch large-scale HPT jobs on Vertex AI with code examples. Additionally, we discussed some best practices to get the best out of both HPT and NAS.

After reading this chapter, you should have a fair understanding of the term "model optimization" and its importance while developing ML applications. Additionally, you should now be confident about quickly setting up small to large-scale hyperparameter tuning experiments with the help of Vertex AI tooling on Google Cloud. You should also have a fair understanding of NAS, its differences from HPT, and the best practices for setting up a NAS job.

Now that we understand the importance and common methods of model optimization techniques, we are in good shape to develop high-quality models. Next, let's learn about how to deploy these models so that they can be consumed by downstream applications.

# 10

# Vertex AI Deployment and Automation Tools – Orchestration through Managed Kubeflow Pipelines

In a typical **machine learning** (**ML**) solution, we often have lots of applications and services as part of the end-to-end workflow. If we try to stitch these services and applications together using some custom scripts with cron jobs, it becomes super tricky to manage the workflows. Thus, it becomes important to make use of some orchestration services to carefully manage, scale, and monitor complex workflows. Orchestration is the process of stitching multiple applications or services together to build an end-to-end solution workflow. Google Cloud provides multiple orchestration services, such as Cloud Scheduler, Workflows, and Cloud Composer, to manage complex workflows at scale. Cloud Scheduler is ideal for single, repetitive tasks, Workflows is more suitable for complex multi-service orchestration, and Cloud Composer is ideal for data-driven workloads.

ML workflows have a lot of steps, from data preparation to model training, evaluation, and more. On top of that, monitoring and version tracking become even more challenging. In this chapter, we will learn about GCP tooling for orchestrating ML workflows effectively. The main topics covered in this chapter are as follows:

- Orchestrating ML workflows using Vertex AI Pipelines (managed Kubeflow pipelines)

- Orchestrating ML workflows using Cloud Composer (managed Airflow)

- Vertex AI Pipelines versus Cloud Composer

- Getting predictions on Vertex AI

- Managing deployed models on Vertex AI

# Technical requirements

The code examples shown in this chapter can be found in the following GitHub repo: `https://github.com/PacktPublishing/The-Definitive-Guide-to-Google-Vertex-AI/tree/main/Chapter10`

# Orchestrating ML workflows using Vertex AI Pipelines (managed Kubeflow pipelines)

ML solutions are complex and involve lots of steps, including data preparation, feature engineering, model selection, model training, testing, evaluation, and deployment. On top of these, it is really important to track and version control lots of aspects related to the ML model while in production. Vertex AI Pipelines on GCP lets us codify our ML workflows in such a way that they are easily composable, shareable, and reproducible. Vertex AI Pipelines can run Kubeflow as well as **TensorFlow Extended** (**TFX**)-based ML pipelines in a fully managed way. In this section, we will learn about developing Kubeflow pipelines for ML development as Vertex AI Pipelines.

Kubeflow is a Kubernetes-native solution that simplifies the orchestration of ML pipelines and makes experimentation easy and reproducible. Also, the pipelines are sharable. It comes with framework support for things such as execution monitoring, workflow scheduling, metadata logging, and versioning. A Kubeflow pipeline is a description of an ML workflow that combines multiple small components of the workflow into a **directed acyclic graph** (**DAG**). Behind the scenes, it runs the pipeline components on containers, which provide portability, reproducibility, and encapsulation. Each pipeline component is one step in the ML workflow that does a specific task. The output of one component may become the input of another component and so forth. Each pipeline component is made up of code, packaged as a Docker image that performs one step in the pipeline and runs on one or more Kubernetes Pods. Kubeflow pipelines can be leveraged for ETL and CI/CD tasks but they are more popularly used to run ML workflows.

The Vertex AI SDK lets us create and upload Kubeflow pipelines programmatically from within the Jupyter Notebook itself, but we can also use the console UI to work on pipelines. The Vertex AI UI lets us visualize the pipeline execution graph. It also lets us track, monitor, and compare different pipeline executions.

## Developing Vertex AI Pipeline using Python

In this section, we will develop and launch a simple Kubeflow-based Vertex Pipeline using the Vertex AI SDK within a Jupyter Notebook. In this example, we will work on an open source wine quality dataset. Let's get started!

Open a Jupyter Notebook and install some useful libraries:

```
!pip3 install google-cloud-aiplatform
!pip3 install kfp --upgrade
!pip install google_cloud_pipeline_components
```

In a new cell, import useful libraries for Vertex Pipeline development:

```
from typing import NamedTuple
import typing
import pandas as pd
from kfp.v2 import dsl
from kfp.v2.dsl import (Artifact, Dataset, Input, Model, Output,
Metrics, ClassificationMetrics, component, OutputPath, InputPath)

from kfp.v2 import compiler
from google.cloud import bigquery
from google.cloud import aiplatform
from google.cloud.aiplatform import pipeline_jobs
from google_cloud_pipeline_components import aiplatform as gcc_aip
```

Create a timestamp variable. It will be useful in creating unique names for pipeline objects:

```
from datetime import datetime
TIMESTAMP =datetime.now().strftime("%Y%m%d%H%M%S")
```

Now, we will set some project-related configurations, such as `project_id`, region, staging bucket, and service account:

```
PROJECT_ID='417xxxxxxx97'
REGION='us-west2'
SERVICE_ACCOUNT='417xxxxxxx97-compute@developer.gserviceaccount.com'
BUCKET_URI='gs://my-training-artifacts'
```

In this section we will use the Wine Quality dataset. The Wine Quality dataset was created by Cortez,Paulo, Cerdeira,A., Almeida,F., Matos,T., and Reis,J.. (2009). You can check it out at OR You can download the dataset from the following link: `https://doi.org/10.24432/C56S3T`. (`UCI Machine Learning Repository.`)

Next, we load and check the wine quality dataset in a notebook cell to understand the data and columns:

```
df_wine = pd.read_csv("http://archive.ics.uci.edu/ml/machine-learning-
databases/wine-quality/winequality-white.csv", delimiter=";")
df_wine.head()
```

The output of this snippet is shown in *Figure 10.1*.

| | fixed acidity | volatile acidity | citric acid | residual sugar | chlorides | free sulfur dioxide | total sulfur dioxide | density | pH | sulphates | alcohol | quality |
|---|---|---|---|---|---|---|---|---|---|---|---|---|
| 0 | 7.0 | 0.27 | 0.36 | 20.7 | 0.045 | 45.0 | 170.0 | 1.0010 | 3.00 | 0.45 | 8.8 | 6 |
| 1 | 6.3 | 0.30 | 0.34 | 1.6 | 0.049 | 14.0 | 132.0 | 0.9940 | 3.30 | 0.49 | 9.5 | 6 |
| 2 | 8.1 | 0.28 | 0.40 | 6.9 | 0.050 | 30.0 | 97.0 | 0.9951 | 3.26 | 0.44 | 10.1 | 6 |
| 3 | 7.2 | 0.23 | 0.32 | 8.5 | 0.058 | 47.0 | 186.0 | 0.9956 | 3.19 | 0.40 | 9.9 | 6 |
| 4 | 7.2 | 0.23 | 0.32 | 8.5 | 0.058 | 47.0 | 186.0 | 0.9956 | 3.19 | 0.40 | 9.9 | 6 |

Figure 10.1 – Overview of the wine quality dataset

Here is a quick overview of the feature columns:

- `volatile acidity`: The `volatile acidity` column represents the amount of gaseous acids

- `fixed acidity`: The amount of fixed acids found in wine, which can be tartaric, succinic, citric, malic, and so on

- `residual sugar`: This column represents the amount of sugar left after the fermentation of wine

- `citric acid`: The amount of citric acid, which is naturally found in fruits

- `chlorides`: The amount of salt in the wine

- `free sulfur dioxide`: Sulpher dioxide, or $SO_2$, prevents wine oxidation and spoilage

- `total sulfur dioxide`: The total amount of $SO_2$ in a wine

- pH: pH is used for checking acidity in a wine

- `density`: Represents the density of the wine

- `sulphates`: Sulphates help preserve the freshness of wine and also protect it from oxidation and bacteria

- `alcohol`: The percentage of alcohol present in the wine

The idea is to predict the wine quality given all the preceding parameters. We will convert it into a classification problem and call a wine *best quality* if its quality indicator value is >=7.

## Pipeline components

In this exercise, we will define four pipeline components for our task:

- Data loading component
- Model training component
- Model evaluation component
- Model deploying component

Here, the first component loads the data and the second component uses that data to train a model. The third component evaluates the trained model on the test dataset. The fourth component automatically deploys the trained model as a Vertex AI endpoint. We will put a condition on automatic model deployment, such as if model ROC >= 0.8, then deploy the model, otherwise don't.

Now, let's define these components one by one. The following is the first component that loads and splits the data into training and testing partitions.

To create a Kubeflow component, we can wrap our function with an @component decorator. Here, we can define the base image, and also the dependencies to install:

> **Note**
> In a real project or production pipeline, it is advisable to write package versions along with their names to avoid any version realated conflicts.

```
@component(
    packages_to_install=["pandas", "pyarrow", "scikit-learn==1.0.0"],
    base_image="python:3.9",
    output_component_file="load_data_component.yaml"
)
```

Here, we define the function that loads and splits the data into train and test sets:

```
def get_wine_data(
    url: str,
    dataset_train: Output[Dataset],
    dataset_test: Output[Dataset]
):
    import pandas as pd
    import numpy as np
    from sklearn.model_selection import train_test_split as tts

    df_wine = pd.read_csv(url, delimiter=";")
    df_wine['best_quality'] = [1 if x>=7 else 0 for x in df_wine.
quality]
    df_wine['target'] = df_wine.best_quality
    df_wine = df_wine.drop(
        ['quality', 'total sulfur dioxide', 'best_quality'],
        axis=1,
    )
```

We will keep about 30% of the data for testing and the remaining for training and save them as CSV files:

```
train, test = tts(df_wine, test_size=0.3)
train.to_csv(
    dataset_train.path + ".csv",
    index=False,
    encoding='utf-8-sig',
)
test.to_csv(
    dataset_test.path + ".csv",
    index=False,
    encoding='utf-8-sig',
)
```

To define a component, we can wrap our Python functions with an @component decorator. It allows us to pass the base image path, packages to install, and a YAML file path if we wish to write the component into a file. The YAML file definition of a component makes it portable and reusable. We can simply create a YAML file with the component definition and load this component anywhere in the project. Note that we can use our custom container image with all our custom dependencies as well.

The first component essentially loads the wine quality dataset table, creates the binary classification output, as discussed previously, drops unnecessary columns, and finally divides it into train and test files. Here, the train and test dataset files are output artifacts of this component that can be reused by subsequently running components.

Now, let's define the second component, which trains a random forest classifier over the training dataset generated by the first component.

The first step is the decorator, with dependencies:

```
@component(
    packages_to_install = [
        "pandas",
        "scikit-learn"
    ],
    base_image="python:3.9",
    output_component_file="model_training_component.yml",
)
```

Next, we define our training function, which fits our model on training data and saves it as a Pickle file. Here, our output artifact would be a model and we can associate it with some metadata as well, as shown in the following function. Inside this function, we can associate the model artifact with metadata by putting the metadata key and value within model.metadata dictionary.

```
def train_winequality(
    dataset:  Input[Dataset],
```

```
        model: Output[Model],
    ):
        from sklearn.ensemble import RandomForestClassifier
        import pandas as pd
        import pickle

        data = pd.read_csv(dataset.path+".csv")
        model_rf = RandomForestClassifier(n_estimators=10)
        model_rf.fit(
            data.drop(columns=["target"]),
            data.target,
        )
        model.metadata["framework"] = "RF"
        file_name = model.path + f".pkl"
        with open(file_name, 'wb') as file:
            pickle.dump(model_rf, file)
```

This component trains a random forest classifier model on the training dataset and saves the model as a Pickle file.

Next, let's define the third component for model evaluation. We start with the @component decorator:

```
@component(
    packages_to_install = [
        "pandas",
        "scikit-learn"
    ],
    base_image="python:3.9",
    output_component_file="model_evaluation_component.yml",
)
```

Now, we define the actual Python function for model evaluation:

```
def winequality_evaluation(
    test_set:  Input[Dataset],
    rf_winequality_model: Input[Model],
    thresholds_dict_str: str,
    metrics: Output[ClassificationMetrics],
    kpi: Output[Metrics]
) -> NamedTuple("output", [("deploy", str)]):

    from sklearn.ensemble import RandomForestClassifier
    import pandas as pd
    import logging
    import pickle
```

```
    from sklearn.metrics import roc_curve, confusion_matrix, accuracy_
score
    import json
    import typing
```

Here is a small function that controls the deployment of the model. We only deploy a new model if its accuracy is above a certain threshold:

```
def threshold_check(val1, val2):
    cond = "false"
    if val1 >= val2 :
        cond = "true"
    return cond


data = pd.read_csv(test_set.path+".csv")
model = RandomForestClassifier()
file_name = rf_winequality_model.path + ".pkl"
with open(file_name, 'rb') as file:
    model = pickle.load(file)


y_test = data.drop(columns=["target"])
y_target=data.target
y_pred = model.predict(y_test)
```

Now that we have the model outputs, we can calculate accuracy scores and roc_curve, and log them as metadata:

```
y_scores =  model.predict_proba(
    data.drop(columns=["target"])
)[:, 1]
fpr, tpr, thresholds = roc_curve(
     y_true=data.target.to_numpy(),
     y_score=y_scores, pos_label=True
)
metrics.log_roc_curve(
    fpr.tolist(),
    tpr.tolist(),
    thresholds.tolist()
)


metrics.log_confusion_matrix(
    ["False", "True"],
    confusion_matrix(
        data.target, y_pred
```

```
    ).tolist(),
)
```

Finally, we check the model accuracy and see whether it satisfies the deployment condition. We return the deployment condition flag from here:

```
accuracy = accuracy_score(data.target, y_pred.round())
thresholds_dict = json.loads(thresholds_dict_str)
rf_winequality_model.metadata["accuracy"] = float(accuracy)
kpi.log_metric("accuracy", float(accuracy))
deploy = threshold_check(float(accuracy), int(thresholds_
dict['roc']))
return (deploy,)
```

This component uses the outputs of component 1 (test dataset) and component 2 (trained model) as input and performs model evaluation. This component performs the following operations:

- Loads the test dataset

- Loads the trained model from a Pickle file

- Logs the ROC curve and confusion matrix as an output artifact

- Checks whether the model accuracy is greater than the threshold

Finally, we define the model deployment component. This component automatically deploys the trained model as a Vertex AI endpoint if the deployment condition is true:

```
@component(
    packages_to_install=["google-cloud-aiplatform", "scikit-
learn",  "kfp"],
    base_image="python:3.9",
    output_component_file="model_winequality_component.yml"
)
```

Next, we define the function that deploys the wine quality model when the deployment condition is true. This function will be wrapped around by the previously defined @component decorator so that we can later use it in the final pipeline definition:

```
def deploy_winequality(
    model: Input[Model],
    project: str,
    region: str,
    serving_container_image_uri : str,
    vertex_endpoint: Output[Artifact],
    vertex_model: Output[Model]
):
```

```
from google.cloud import aiplatform
aiplatform.init(project=project, location=region)

DISPLAY_NAME  = "winequality"
MODEL_NAME = "winequality-rf"
ENDPOINT_NAME = "winequality_endpoint"
```

Here, we define a function to create the endpoint for our model so that we can use it for inference:

```
def create_endpoint():
    endpoints = aiplatform.Endpoint.list(
    filter='display_name="{}"'.format(ENDPOINT_NAME),
    order_by='create_time desc',
    project=project,
    location=region,
    )
    if len(endpoints) > 0:
        endpoint = endpoints[0]  # most recently created
    else:
        endpoint = aiplatform.Endpoint.create(
        display_name=ENDPOINT_NAME, project=project,
location=region
        )
    endpoint = create_endpoint()
```

Here, we import our saved model programmatically:

```
#Import a model programmatically
model_upload = aiplatform.Model.upload(
    display_name = DISPLAY_NAME,
    artifact_uri = model.uri.replace("model", ""),
    serving_container_image_uri =  serving_container_image_uri,
    serving_container_health_route=f"/v1/models/{MODEL_NAME}",
    serving_container_predict_route=f"/v1/models/{MODEL_
NAME}:predict",
    serving_container_environment_variables={
    "MODEL_NAME": MODEL_NAME,
    },
    )
```

Finally, we deploy the uploaded model on the desired machine type with the desired traffic split:

```
model_deploy = model_upload.deploy(
    machine_type="n1-standard-4",
    endpoint=endpoint,
```

```
        traffic_split={"0": 100},
        deployed_model_display_name=DISPLAY_NAME,
    )


    # Save data to the output params
    vertex_model.uri = model_deploy.resource_name
```

Now that the core components of our pipeline are ready, we can go ahead and define our Vertex Pipeline.

First, we need to provide a unique name for our pipe:

```
DISPLAY_NAME = 'pipeline-winequality job{}'.format(TIMESTAMP)
```

Pipeline definition is the part where we stitch these components together to define our ML workflow (or execution graph). Here, we can control which components run first and the output of which component should be fed to another component. The following scripts define a simple pipeline for our experiment.

We can use the @dsl.pipeline decorator to define a Kubeflow pipeline. We can pass here a **Google Cloud Storage (GCS)** location where it will store any execution-related artifacts. This GCS location is passed through the pipeline_root parameter inside the decorator, as shown in the following code:

```
@dsl.pipeline(
    pipeline_root=BUCKET_URI,
    name="pipeline-winequality",
)
def pipeline(
    url: str = "http://archive.ics.uci.edu/ml/machine-learning-
databases/wine-quality/winequality-white.csv",
    project: str = PROJECT_ID,
    region: str = REGION,
    display_name: str = DISPLAY_NAME,
    api_endpoint: str = REGION+"-aiplatform.googleapis.com",
    thresholds_dict_str: str = '{"roc":0.8}',
    serving_container_image_uri: str = "us-docker.pkg.dev/vertex-ai/
prediction/sklearn-cpu.0-24:latest"
    ):
```

We can create the execution DAG here and define the order of execution for our predefined components. Some components can be dependent, where the output of one component is the input for another. Dependent components execute sequentially, while independent ones can be executed in parallel:

```
    # adding first component
    data_op = get_wine_data(url)
    # second component uses output of first component as input
    train_model_op = train_winequality(data_op.outputs["dataset_
```

```
train"])
    # add third component (uses outputs of comp1 and comp2 as input)
    model_evaluation_op = winequality_evaluation(
        test_set=data_op.outputs["dataset_test"],
        rf_winequality_model=train_model_op.outputs["model"],
        # We deploy the model only if the model performance is above
the threshold
        thresholds_dict_str = thresholds_dict_str,
    )
```

Here is our condition that decides whether to deploy this model or not:

```
# condition to deploy the model
with dsl.Condition(
    model_evaluation_op.outputs["deploy"]=="true",
    name="deploy-winequality",
):
    deploy_model_op = deploy_winequality(
    model=train_model_op.outputs['model'],
    project=project,
    region=region,
    serving_container_image_uri = serving_container_image_uri,
    )
```

Here, we use the @dsl.pipeline decorator to define our pipeline. Note that in the preceding definition, the first three components are simple, but the fourth component has been defined using dsl.Condition(). We only run the model deployment component if this condition is satisfied. So, this is how we can control when to deploy the model. If our model meets the business criteria, we can choose to auto-deploy it.

Next, we can compile our pipeline:

```
compiler.Compiler().compile(
    pipeline_func=pipeline,
    package_path='ml_winequality.json',
)
```

Finally, we can submit our pipeline job to Vertex AI:

```
pipeline_job = pipeline_jobs.PipelineJob(
    display_name="winequality-pipeline",
    template_path="ml_winequality.json",
    enable_caching=False,
    location=REGION,
```

```
)

pipeline_job.run()
```

This script will launch our pipeline in Vertex AI. It will also provide us with a console URL to monitor the pipeline job.

We can also locate the pipeline run by going to the **Vertex AI** tab in the console and clicking on the **pipelines** tab. *Figure 10.2* is a screenshot of the execution graph present in Vertex AI for our example job.

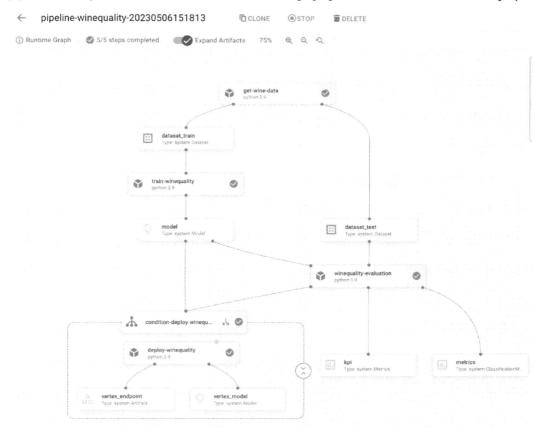

Figure 10.2 – Execution graph of our example Vertex Pipeline from the Google Cloud console UI

As we can see, this execution graph has all four components defined by us. It also has all the artifacts generated by the components. If we click on the **metrics** artifact, we can see the output values in the right pane of the console UI. It looks something similar to *Figure 10.3*.

VIEW LINEAGE

| | |
|---|---|
| **Name** | metrics |
| **Type** | system.ClassificationMetrics |
| **URI** | gs://my-training-artifacts/417812395597/pipeline-winequality-20230506151813/winequality-evaluation_-5446127382979149824/metrics |

## Confusion matrix

 Item counts ⬇

This table shows how often the model classified each label correctly (in blue), and which labels were most often confused for that label (in gray).

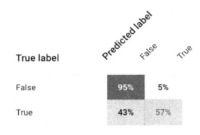

### ROC Curve

Receiver operating characteristic (ROC) curve is a graph showing the performance of a classification model. The curve plots TPR vs. FPR at different classification thresholds.

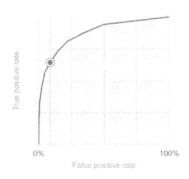

Figure 10.3 – Metadata and artifacts related to our pipeline execution

This is how we can use the Google Cloud console UI to track the execution and metrics of our ML-related workflows. Once we have our pipeline ready, we can also schedule its execution using services such as the native scheduler for Vertex AI Pipelines, Cloud Scheduler (we can define a schedule), Cloud Functions (event-based trigger), and so on.

Now, we have a good understanding of how Kubeflow pipelines can be developed on Google Cloud as Vertex AI Pipelines. We should be able to develop and launch our custom pipelines from scratch now. In the next section, we will learn about Cloud Composer as another solution for workflow orchestration.

# Orchestrating ML workflows using Cloud Composer (managed Airflow)

Cloud Composer is a workflow orchestration service on Google Cloud that is built upon the open source project of Apache Airflow. The key difference is that Composer is fully managed and also integrates with other GCP tooling very easily. With Cloud Composer, we can write, execute, schedule, or monitor our workflows that are also supported across multi-cloud and hybrid environments. Composer pipelines are DAGs that can be easily defined and configured using Python. It comes with a rich library of connectors that let us deploy our workflows instantly with one click. Graphical representations of workflows on the Google Cloud console make monitoring and troubleshooting quite convenient. Automatic synchronization of our DAGs ensures that our jobs always stay on schedule.

Cloud Composer is commonly used by data scientists and data engineers to build complex data pipelines (ETL or ELT pipelines). It can also be used as an orchestrator for ML workflows. Cloud Composer is pretty convenient for data-related workflows as the Apache project comes with hundreds of operators and sensors that make it easy to communicate across multiple cloud environments with very little code. It also lets us define failure handling mechanisms such as sending emails or Slack notifications on pipeline failure.

Now let's understand how to develop Cloud Composer-based pipelines.

## Creating a Cloud Composer environment

We can follow these steps to create a Cloud Composer environment using the Google Cloud console UI:

1. Enable the Cloud Composer API.
2. From the left pane of the console, select **Composer** and click on **Create** to start creating a Composer environment (see *Figure 10.4*).

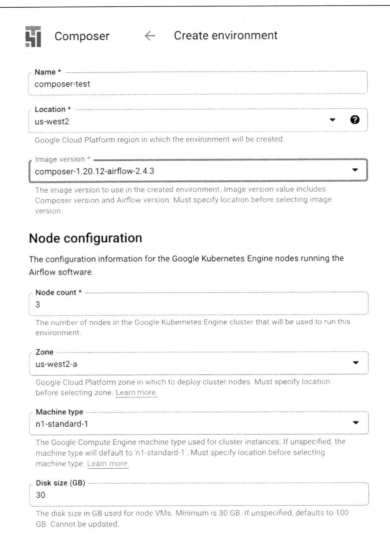

Figure 10.4 – Creating a Composer environment on the Google Cloud console

3.   Click on **Create**. It will take about 15–20 minutes to create the environment. Once it is complete, the environment page will look like the following (see *Figure 10.5*).

Figure 10.5 – Ready-to-use Cloud Composer environment

4.  Click on **Airflow** to see the Airflow web UI. The Airflow web UI is shown in *Figure 10.6*.

Figure 10.6 – Airflow web UI with our workflows

As we can see in the preceding screenshot, there is already one DAG running – **airflow_monitoring**. If we go to the Cloud Storage bucket for Composer, we will find a **dags** folder there, and we will be able to see one `airflow_monitoring.py` file. See *Figure 10.7*.

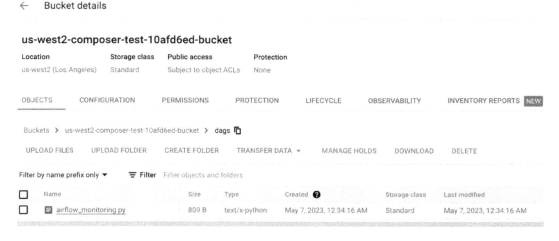

Figure 10.7 – GCS location where we can put our Python-based DAGs for execution

Now that our Composer setup is ready, we can quickly check whether it is working as expected. To test things fast, we will use one demo DAG from the Airflow tutorials and put it inside the **dags** folder of this bucket. If everything is working fine, any DAG that we put inside this bucket should automatically get synced with Airflow.

The following is the code for a demo DAG from the Airflow tutorials:

```
from datetime import timedelta
from textwrap import dedent

# The DAG object; we'll need this to instantiate a DAG
```

```
from airflow import DAG

# Operators; we need this to operate!
from airflow.operators.bash import BashOperator
from airflow.utils.dates import days_ago
# These args will get passed on to each operator
# You can override them on a per-task basis during operator
initialization
```

The following dictionary with some default arguments will be used when creating the operators later. By default, these arguments will be passed to each operator, but we can also override some of these arguments in some operators as per the requirements:

```
default_args = {
    'owner': 'airflow',
    'depends_on_past': False,
    'email': ['airflow@example.com'],
    'email_on_failure': False,
    'email_on_retry': False,
    'retries': 1,
    'retry_delay': timedelta(minutes=5),
    # 'queue': 'bash_queue',
    # 'execution_timeout': timedelta(seconds=300),
    # 'on_failure_callback': some_function,
    # 'on_success_callback': some_other_function,
    # 'on_retry_callback': another_function,
    # 'sla_miss_callback': yet_another_function,
}
```

This is where we define our DAG, with execution steps in the desired or required order:

```
with DAG(
    'composer-test-dag',
    default_args=default_args,
    description='A simple composer DAG',
    schedule_interval=timedelta(days=1),
    start_date=days_ago(2),
    tags=['example'],
) as dag:
```

Here, we define different tasks that our code will be performing:

```
    # t1, t2 and t3 are examples of tasks created by instantiating
operators
    t1 = BashOperator(
```

```
        task_id='print_date',
        bash_command='date',
    )


    t2 = BashOperator(
        task_id='sleep',
        depends_on_past=False,
        bash_command='sleep 5',
        retries=3,
    )
```

You can document your task using the following attributes: doc_md (Markdown), doc (plain text), doc_rst, doc_json, and doc_yaml, which gets rendered on the UI's **Task Instance Details** page:

```
    t1.doc_md = dedent(
    )

    dag.doc_md = __doc__
    dag.doc_md = """a documentation placed anywhere"""
    templated_command = dedent(
        """
    {% for i in range(5) %}
        echo "{{ ds }}"
        echo "{{ macros.ds_add(ds, 7)}}"
        echo "{{ params.my_param }}"
    {% endfor %}
    """
    )
```

Now let's define the t3 task:

```
    t3 = BashOperator(
        task_id='templated',
        depends_on_past=False,
        bash_command=templated_command,
        params={'my_param': 'Parameter I passed in'},
    )
```

Here, we define the execution order of our tasks. t1 needs to be executed before t2 and t3, but t2 and t3 can execute in parallel:

```
    t1 >> [t2, t3]
```

As soon as we upload this `.py` file to a GCS bucket inside the dags folder, Airflow will automatically sync it. If you refresh the Airflow web UI, it should show another DAG, as shown in *Figure 10.8*.

Figure 10.8 – Airflow web UI with all the DAGs that are present in the GCS location

If we are able to see our DAG running in the Airflow UI, it verifies that our installation is working fine. Now, let's open this DAG to check the actual execution graph. It should look something similar to what is shown in *Figure 10.9*.

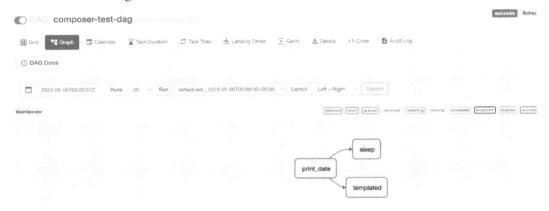

Figure 10.9 – Execution graph of our workflow within the Airflow web UI

Although it is a very simple DAG, it gives an idea of how easy it is to work with Airflow using Cloud Composer. The level of logging and monitoring we get with Cloud Composer is quite amazing. Cloud Composer makes the lives of data engineers really easy so that they can focus on defining complex data pipelines without worrying about infrastructure and Airflow management.

We now have a good idea of how Vertex AI Pipelines and Cloud Composer (managed Airflow service) can be used as an orchestrator for ML workflows. Now let's summarize some of the similarities and differences between these two.

## Vertex AI Pipelines versus Cloud Composer

In this section, we will talk about some of the key similarities and differences between Vertex AI Pipelines and Cloud Composer when it comes to orchestrating ML workflows. Based on this comparison, we can choose the best solution for our next ML project. The following is a list of points that summarize the important aspects of both orchestrators for ML-related tasks:

- Both are easy to use and divide the overall ML workflow into smaller execution units in terms of tasks (Composer) or containerized components (Vertex AI Pipelines).

- Passing data between components is similar, and it may require an intermediate storage system if the data size is large.

- Vertex AI Pipelines have an extensive list of prebuilt components available open source and thus developers can avoid writing a lot of boilerplate code. On the other hand, in the case of a Composer-based pipeline, we need to write the entire workflow.

- Based on the ease of setting up environments, Vertex AI Pipelines is a little bit easier.

- Both run on Kubernetes, but in the case of Vertex AI Pipelines, there is no need to worry about clusters, Pods, and so on.

- Cloud Composer is ideal for data-related tasks. We can also implement ML pipelines as a data task but we lose a lot of ML-related functionalities, such as lineage tracking, metrics, experiment comparisons, and distributed training. These features come out of the box with Vertex AI Pipelines.

- Data engineers might feel more comfortable with Composer pipelines, while ML engineers might be more comfortable with Vertex AI Pipelines.

- In many cases, Vertex AI Pipelines can be cheaper to use as here we pay for what we use. On the other hand, in the case of Composer, some Pods are always running.

- If needed, some of the Vertex AI Pipelines capabilities can be used with Composer as well.

- Working with Vertex AI Pipelines requires zero knowledge about Kubernetes, but with Cloud Composer, it is important to know common aspects of Kubernetes.

After reading these comparison points, we might find it easy to choose the best orchestrator for our next ML use case. Nevertheless, both orchestrators are easy to use and are commonly used across organizations to manage their complex data/ML-related workflows.

Now that we have a good understanding of ML orchestration tools on Google Cloud with their pros and cons, we are ready to start developing production-grade ML pipelines. Next, let's learn how to get predictions on Vertex AI.

## Getting predictions on Vertex AI

In this section, we will learn how to get predictions from our ML models on Vertex AI. Depending on the use case, prediction requests can be of two types – online predictions (real time) and batch predictions. Online predictions are synchronous requests made to a model endpoint. Online predictions are needed by applications that keep requesting outputs for given inputs in a timely manner via an API call in order to update information for end users in near real time. For example, the Google Maps API gives us near real-time traffic updates and requires online prediction requests. Batch predictions, on the other hand, are asynchronous requests. If our use case only requires batch prediction, we might not need to deploy the model to an endpoint as the Vertex AI `batchprediciton` service also allows us to perform batch prediction from a saved model that is present in a GCS location without even needing to create an endpoint. Batch predictions are suitable for use cases where the response is not time sensitive and we can afford to get a delayed response (for example, an e-commerce company may wish to forecast sales for the next six months or so). Using batch predictions, we can make predictions of a large amount of data with just a single request.

## Getting online predictions

We must deploy our model to an endpoint before that model can be used to serve online prediction requests. Model deployment essentially means keeping the model in memory with the required infrastructure (memory and compute) so that it can serve predictions with low latency. We can deploy multiple models to a single endpoint as well as a single model to multiple endpoints based on the use case and scaling requirements.

When you deploy a model using the Vertex AI API, you complete the following steps:

1. Create an endpoint.

2. Get the endpoint ID.

3. Deploy the model to the endpoint.

We can use the following Python sample function to create a Vertex AI endpoint. This function is taken from official documentation (`https://cloud.google.com/vertex-ai/docs/general/deployment#api`):

```
def create_endpoint_sample(
    project: str,
    display_name: str,
    location: str,
):
    aiplatform.init(project=project, location=location)

    endpoint = aiplatform.Endpoint.create(
        display_name=display_name,
        project=project,
        location=location,
    )

    print(endpoint.display_name)
    print(endpoint.resource_name)
    return endpoint
```

The second step is to get the endpoint ID so that we can use it to deploy our model. The following shell command will give us a list of all the endpoints within our project and location. We can filter it with the endpoint name if we have it:

```
gcloud ai endpoints list \
  --region=LOCATION \
  --filter=display_name=ENDPOINT_NAME
```

Now that we have the endpoint ID, we can deploy our model to this endpoint. While deploying the model, we can specify parameters for a number of replicas, the accelerator count, accelerator types, and so on. The following is a sample Python function that can be used to deploy the model to a given endpoint. This sample has been taken from the Google Cloud documentation (https://cloud.google.com/vertex-ai/docs/general/deployment#api):

```
def deploy_model_with_dedicated_resources_sample(
    project,
    location,
    model_name: str,
    machine_type: str,
    endpoint: Optional[aiplatform.Endpoint] = None,
    deployed_model_display_name: Optional[str] = None,
    traffic_percentage: Optional[int] = 0,
    traffic_split: Optional[Dict[str, int]] = None,
    min_replica_count: int = 1,
    max_replica_count: int = 1,
    accelerator_type: Optional[str] = None,
    accelerator_count: Optional[int] = None,
    explanation_metadata: Optional[explain.ExplanationMetadata] =
None,
    explanation_parameters: Optional[explain.ExplanationParameters] =
None,
    metadata: Optional[Sequence[Tuple[str, str]]] = (),
    sync: bool = True,
):
```

Here, we initialize the Vertex AI SDK and deploy our model to an endpoint:

```
    aiplatform.init(project=project, location=location)

    model = aiplatform.Model(model_name=model_name)

    model.deploy(
        endpoint=endpoint, deployed_model_display_name=deployed_model_
display_name,
        traffic_percentage=traffic_percentage,
```

```
        traffic_split=traffic_split,
        machine_type=machine_type,
        min_replica_count=min_replica_count,
        max_replica_count=max_replica_count,
        accelerator_type=accelerator_type,
        accelerator_count=accelerator_count,
        explanation_metadata=explanation_metadata,
        explanation_parameters=explanation_parameters,
        metadata=metadata,
        sync=sync,
    )

    model.wait()

    print(model.display_name)
    print(model.resource_name)
    return model
```

Once our model is deployed to an endpoint, it is ready to serve online predictions. We can now make online prediction requests to this endpoint. See the following sample request:

```
def endpoint_predict_sample(
    project: str, location: str, instances: list, endpoint: str
):
    aiplatform.init(project=project, location=location)

    endpoint = aiplatform.Endpoint(endpoint)

    prediction = endpoint.predict(instances=instances)
    print(prediction)
    return prediction
```

The instances[] object is required and must contain the list of instances to get predictions for. See the following example:

```
{
  "instances": [
    [0.0, 1.1, 2.2],
    [3.3, 4.4, 5.5],
    ...
  ]
}
```

The response body is also similar. It may look something like the following example. This example is not related to the earlier model; it is just for understanding purposes:

```
{
  "predictions": [
    {
      "label": "tree",
      "scores": [0.2, 0.8]
    },
    {
      "label": "bike",
      "scores": [0.85, 0.15]
    }
  ],
  "deployedModelId": 123456789012345678
}
```

The response when there is an error in processing the input looks as follows:

```
{"error": "Divide by zero"}
```

We now have a good idea of how to get online predictions using Vertex AI endpoints. But not every use case requires on-demand or online predictions. There are times when we want to make predictions on a large amount of data but the results are not immediately required. In such cases, we can utilize batch predictions. Let's discuss more about getting batch predictions using Vertex AI.

## Getting batch predictions

As discussed before, batch prediction requests are asynchronous and do not require a model to be deployed to an endpoint all the time. To make a batch prediction request, we specify an input source and an output location (either Cloud Storage or BigQuery), where Vertex AI stores prediction results. The input source location must contain our input instances in one of the accepted formats: TFRecord, JSON Lines, CSV, BigQuery, and so on.

TFRecord input instances may look something like the following example:

```
{"instances": [
    { "b64": "b64EncodedASCIIString" },
    { "b64": "b64EncodedASCIIString" }
]}
```

Batch prediction can be requested through Vertex AI API programatically or also with Google Cloud console UI. As we can pass lots of data to batch prediction requests, they may take a long time to complete depending upon the size of data and model.

A sample batch prediction request using the Vertex AI API with Python may look something like the following Python function. This sample code has been taken from the official documentation (https://cloud.google.com/vertex-ai/docs/predictions/get-batch-predictions):

```
def create_batch_prediction_job_dedicated_resources_sample(
    project: str,
    location: str,
    model_resource_name: str,
    job_display_name: str,
    gcs_source: Union[str, Sequence[str]],
    gcs_destination: str,
    instances_format: str = "jsonl",
    machine_type: str = "n1-standard-2",
    accelerator_count: int = 1,
    accelerator_type: Union[str, aiplatform_v1.AcceleratorType] =
"NVIDIA_TESLA_K80",
    starting_replica_count: int = 1,
    max_replica_count: int = 1,
    sync: bool = True,
):
```

Here, we initialize the Vertex AI SDK and call batch predictions on our deployed model:

```
aiplatform.init(project=project, location=location)

my_model = aiplatform.Model(model_resource_name)

batch_prediction_job = my_model.batch_predict(
    job_display_name=job_display_name,
    gcs_source=gcs_source,
    gcs_destination_prefix=gcs_destination,
    instances_format=instances_format,
    machine_type=machine_type,
    accelerator_count=accelerator_count,
    accelerator_type=accelerator_type,
    starting_replica_count=starting_replica_count,
    max_replica_count=max_replica_count,
    sync=sync,
)

batch_prediction_job.wait()

print(batch_prediction_job.display_name)
```

```
    print(batch_prediction_job.resource_name)
    print(batch_prediction_job.state)
    return batch_prediction_job
```

Once the batch prediction request is complete, the output is saved in the specified Cloud Storage or BigQuery location.

A jsonl output file might look something like the following example output:

```
{ "instance": [1, 2, 3, 4], "prediction": [0.1,0.9] }
{ "instance": [5, 6, 7, 8], "prediction": [0.7,0.3] }
```

We now have a fair idea of how online and batch prediction work on Vertex AI. The idea of separating batch prediction from online prediction (eliminating the need for deployment) saves a lot of resources and costs. Next, let's discuss some important considerations related to deployed models on Google Vertex AI.

# Managing deployed models on Vertex AI

When we deploy an ML model to an endpoint, we associate it with physical resources (compute) so that it can serve online predictions at low latency. Depending on the requirements, we might want to deploy multiple models to a single endpoint or a single model to multiple endpoints as well. Let's learn about these two scenarios.

## Multiple models – single endpoint

Suppose we already have one model deployed to an endpoint in production and we have found some interesting ideas to improve that model. Now, suppose we have already trained an improved model that we want to deploy but we also don't want to make any sudden changes to our application. In this situation, we can add our latest model to the existing endpoint and start serving a very small percentage of traffic with the new model. If everything looks great, we can gradually increase the traffic until it is serving the full 100% of the traffic.

## Single model – multiple endpoints

This is useful when we want to deploy our model with different resources for different application environments, such as testing and production. Secondly, if one of our applications has high-performance needs, we can serve it using an endpoint with high-performance machines, while we can serve other applications with lower-performance machines to optimize operationalization costs.

## Compute resources and scaling

Vertex AI allocates compute nodes to handle online and batch predictions. When we deploy our ML model to an endpoint, we can customize the type of virtual machines to be used for serving the model. We can choose accelerators such as GPUs or TPUs if needed. A machine configuration with more computing resources can serve predictions with lower latency, hence handling more prediction requests at the same time. But such a machine will cost more than a machine with low compute resources. Thus, it is important to choose the best-suited machine depending on the use case and requirements.

When we deploy a model for online predictions, we can also configure a prediction node to automatically scale. But the prediction nodes for batch prediction do not automatically scale. By default, if we deploy a model with or without dedicated GPU resources, Vertex AI will automatically scale the number of replicas up or down so that CPU or GPU usage (whichever is higher) matches the default 60% target value. Given these conditions, Vertex AI will scale up, even if this may not have been needed to achieve **queries per second** (**QPS**) and latency targets. We can monitor the endpoint to track metrics such as CPU and accelerator usage, the number of requests, and latency, as well as the current and target number of replicas.

To determine the ideal machine type for a prediction container from a cost perspective, we can deploy it to a virtual machine instance and benchmark the instance by making prediction requests until the virtual machine hits about 90% of the CPU usage. By doing this experiment a few times on different machines, we can identify the cost of the prediction service based on the QPS values.

## Summary

In this chapter, we have learned about two popular ML workflow orchestration tools – Vertex AI Pipelines (managed Kubeflow) and Cloud Composer (managed Airflow). We have also implemented a Vertex Pipeline for an example use case, and similarly, we have also developed and executed an example DAG with Cloud Composer. Both Vertex AI Pipelines and Cloud Composer are managed services on GCP and make it really easy to set up and launch complex ML and data-related workflows. Finally, we have learned about getting online and batch predictions on Vertex AI for our custom models, including some best practices related to model deployments.

After reading this chapter, you should have a good understanding of different ways of carrying out ML workflow orchestration on GCP and their similarities and differences. Now, you should be able to write your own ML workflows and orchestrate them on GCP via either Vertex AI Pipelines or Cloud Composer. Finally, you should also be confident in getting online and batch predictions using Vertex AI.

Now that we have a good understanding of deploying ML models on GCP, and also orchestrating ML workflows, we can start developing production-grade pipelines for different use cases. Along similar lines, we will learn about some ML governance best practices and tools in the upcoming chapter.

# 11

# MLOps Governance with Vertex AI

In the rapidly evolving digital era, the successful implementation of **machine learning** (**ML**) solutions is not just about creating sophisticated models that can predict outcomes accurately for complex use cases. While this is undoubtedly essential, the proficient management and governance of **artificial intelligence** (**AI**)/**ML operations** (**MLOps**) is equally important. This is especially important in an enterprise setting where companies have to ensure they adhere to several internal policies and regulatory compliance requirements. This chapter delves into different MLOps governance components and how you can utilize features available within Google Cloud to implement them.

MLOps governance revolves around instituting a structured approach to managing and optimizing the various moving parts of ML operations. It encompasses the processes, tools, and guidelines that ensure the smooth functioning of ML projects, all while complying with required policies and regulations.

In this chapter, we will cover MLOps governance on Google Cloud, with a focus on the following key areas:

- Understanding MLOps governance
- Case studies of MLOps governance
- Implementing MLOps governance on Google Cloud

By the end of this chapter, our goal is to equip you with a comprehensive understanding of MLOps governance, its implementation on Google Cloud, and its critical role in maintaining successful, scalable, and compliant ML operations.

# What is MLOps governance and what are its key components?

MLOps refers to the discipline that combines ML, data science, and DevOps principles to manage the life cycle of ML models efficiently. The goal of MLOps is to create a streamlined pipeline for developing, deploying, and maintaining ML models, ensuring that these models provide reliable and consistent results. However, the implementation and management of such a practice require a governing framework to ensure adherence to best practices and standards. This governing framework is what we refer to as MLOps governance.

MLOps governance is an essential, yet often overlooked, aspect of implementing and managing ML models within an organization. It encapsulates a comprehensive set of rules, procedures, and guidelines aimed at overseeing the ML models throughout their life cycle. This governance plays a pivotal role in ensuring that the MLOps pipeline operates smoothly and ethically, mitigating any risks associated with ML model deployment and usage.

The primary focus of MLOps governance is to create a reliable, transparent, and accountable ML system within an organization. This involves overseeing aspects such as data handling, model development, model deployment, model monitoring, and model auditing and can be broken down into two key facets: **data governance** and **model governance**.

## Data governance

ML models are only as good as the data they are trained on. In MLOps governance, data handling refers to the governance of how data is collected, stored, processed, and used. It entails ensuring the quality and relevance of data, preserving data privacy, and complying with relevant regulations. This guarantees that the data that's used for model training is not only of high quality but also ethically sourced and used.

## Model governance

Model governance comprises the following components:

- **Model deployment management**: Overseeing model deployment involves ensuring that the model is correctly integrated into the organization's system and that it operates as expected. It also involves checking that the model doesn't inadvertently cause any harmful outcomes, such as biased results or privacy violations.

- **Model auditing**: MLOps governance ensures that there is a systematic review of the ML models in terms of their performance, ethical implications, and overall impact on the organization. Model auditing is essential to maintain transparency and accountability, particularly in scenarios where the model's predictions significantly influence business decisions or user experiences.

- **Model monitoring**: Once the model has been deployed, MLOps governance requires that it be monitored continuously for any changes in its performance. This includes tracking the model's accuracy, detecting data drift, and making sure the model continues to deliver reliable predictions.

MLOps governance is not a one-size-fits-all practice; it needs to be tailored to the specific needs and circumstances of each organization. This might involve customizing the governance based on the nature of the data being handled, the type of ML models being used, the specific applications of these models, and the broader regulatory landscape.

To summarize, MLOps governance is a critical component of any organization that employs ML models. By establishing robust MLOps governance, organizations can ensure that their MLOps practices are not just effective but also ethical, transparent, and compliant.

# Enterprise scenarios that highlight the importance of MLOps governance

To understand the importance of MLOps governance, let's go through some real-world scenarios that highlight this.

## Scenario 1 – limiting bias in AI solutions

Consider a financial services firm deploying a suite of ML models to predict credit risk. A large firm in the finance sector would have an array of internal policies around the data access, usage, and risk assessment of predictive models that its ML solutions will need to adhere to. This could range from limits on what data can be used for such purposes to who can access the model's outputs. It would also be obligated to follow several regulatory requirements, such as preventing bias against protected classes in its decision-making models. For example, a bank would need to ensure that its decision-making process around loan approval is not biased based on race or gender. Even if the regulators can't decipher the underlying ML models, they can conduct statistical analysis to detect whether there is a significant correlation between loan approvals and factors such as race. If a bank is found to be biased in its decision-making, besides being hit with substantial penalties by the regulators, it would also have a major public relations disaster on its hands. So the bank needs to build checks and balances in their ML development life cycle to flag any such issues in their models under development and prevent such models from ever reaching production environments.

## Scenario 2 – the need to constantly monitor shifts in feature distributions

Consider a scenario where an online e-commerce giant makes extensive use of AI to provide personalized recommendations to its retail customers. It has a set of models that seem to be working well in production. Now, the retailer is making a big marketing push to acquire users in additional regions that have been underrepresented in its customer base so far. As the influx of customers from new regions starts to grow, the retailer's business development team notices a sharp decline in its click-through rate and revenue per user session based on the new monthly sales analytics report. When its analysts dig into the possible causes, they realize that the age distribution of users from the new regions is significantly different from the age distribution of the customers from the existing regions. This type of shift in feature/data distribution is known as **data drift** in MLOps parlance and can have a significant impact on user experience and, ultimately, the company's bottom line. Although we are considering a hypothetical scenario where the company's expansion into additional regions is causing a shift in data, this can happen due to several different scenarios, including, but not limited to, a shift in marketing strategy, a change in the economy, and a change in product offerings.

So, it's important to have checks in place to catch such material changes in inference input data early so that the data science team can mitigate its impact by either building newer models with more recent data or building more targeted models.

## Scenario 3 – the need to monitor costs

With great power comes great responsibility. Just like any other scalable technology in the cloud, there is a possibility of your team running up a huge bill if the resources are not planned properly, and proper budgets and limits are not set in the Google Cloud projects as safeguards. Consider a situation where a data scientist spins up a Vertex AI Workbench environment with an expensive GPU attached to the node for a quick experiment but then forgets to shut down the machine. Another similar scenario would be where someone tries to schedule an MLOps pipeline to run once a month with an extremely large GPU cluster but mistakenly configures it to run once a day, thereby making the cost 30x what it should have been. One or two such mistakes by themselves might not break the bank for a typical mid-size company but you can imagine how such costs can quickly add up, especially in large, distributed teams where no single person has full context of whether a training job running on a $10k/month cluster for last 3 days is an actual experiment being tracked or whether it's just a mistake. So, it's extremely important to set up cost management policies and, more importantly, automated controls that would limit the usage of specific resources on **Google Cloud Platform** (**GCP**).

## Scenario 4 – monitoring how the training data is sourced

Although data controls at the source of the data would primarily be handled by the data owners, AI product/solution leaders need to be cognizant of where they are sourcing their data from. If the data that's being used to train the models is later discovered to be unlicensed or coming from sources with questionable data quality, it can lead to a significant amount of wasted resources, both in terms of infrastructure cost and personnel overhead.

Now, let's look at the different tools and features available within Vertex AI to help you implement MLOps governance across your ML solutions.

# Tools in Vertex AI that can help with governance

Vertex AI offers several tools to help with ML solution governance and monitoring that you can utilize to implement and track your organization's standard governance policies and more generic governance best practices. Please keep in mind that for many of the governance policies, especially the ones around security and cost management, you will need to use tools outside of Vertex AI. For example, to set up monthly cost limits and budgets, you will need to use GCP's native billing tools.

Let's walk through the details of the different tools within Vertex AI that can be used as part of MLOps governance processes:

## Model Registry

Vertex AI Model Registry provides a centralized, organized, and secure location for managing all ML models within an organization. This facilitates seamless and efficient ML operations, from development and validation to deployment and monitoring:

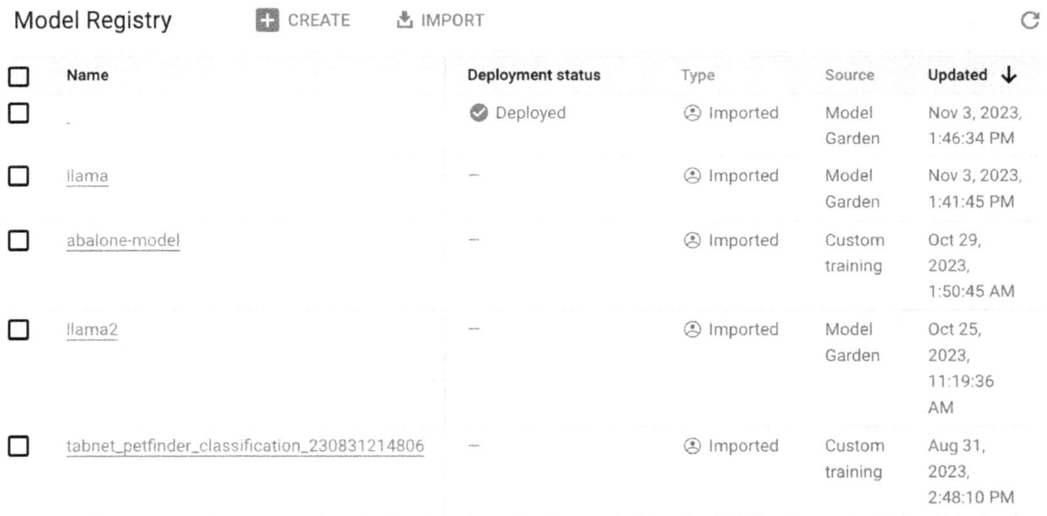

| Model Registry | CREATE | IMPORT | | | | |
|---|---|---|---|---|---|---|
| ☐ **Name** | | | **Deployment status** | Type | Source | **Updated ↓** |
| ☐ . | | | ✔ Deployed | ⊕ Imported | Model Garden | Nov 3, 2023, 1:46:34 PM |
| ☐ llama | | | – | ⊕ Imported | Model Garden | Nov 3, 2023, 1:41:45 PM |
| ☐ abalone-model | | | – | ⊕ Imported | Custom training | Oct 29, 2023, 1:50:45 AM |
| ☐ llama2 | | | – | ⊕ Imported | Model Garden | Oct 25, 2023, 11:19:36 AM |
| ☐ tabnet_petfinder_classification_230831214806 | | | – | ⊕ Imported | Custom training | Aug 31, 2023, 2:48:10 PM |

Figure 11.1 – Vertex AI Model Registry

Acting as a central hub for managing your ML models' life cycles, Vertex AI Model Registry offers a bird's-eye view of your models, thus enabling a more organized and efficient method of tracking and training new model versions. It serves as an access point from where you can deploy your preferred model version to an endpoint, either directly or by employing aliases.

Vertex AI Model Registry extends its support to custom models across all AutoML data types – be it text, tabular data, images, or videos. Moreover, it can incorporate BigQuery ML models, which means that if you have models that have undergone training via BigQuery ML, you can easily register them within Vertex AI Model Registry.

Navigating to the model version details page, you're provided with numerous options: you can evaluate a model, deploy it to an endpoint, set up batch prediction, and inspect specific details related to the model. With its user-friendly and streamlined interface, Vertex AI Model Registry simplifies how you can manage and deploy your optimal models to a production environment.

Let's explore how Vertex AI Model Registry contributes to ML governance:

- **Centralized repository for models**: Model Registry provides a single location where all models in the organization are stored. This centralized repository makes it easy for data scientists, ML engineers, and DevOps teams to store, access, and manage models. It also fosters cross-functional visibility and collaboration, which are essential elements in maintaining a robust governance framework.

- **Version control and model lineage**: Every time a new model is trained or an existing model is updated, a new version is created in Model Registry. It maintains a history of all versions of a model, enabling easy tracking and comparison of different versions and ensuring that any updates or modifications are adequately logged and accounted for:

Figure 11.2 – Vertex AI Model Registry (version view)

- **Model metadata management**: In combination with the Metadata Store, it can help record the model's lineage, providing information about the datasets, model parameters, and training pipelines that are used to build each version of the model. This lineage information is invaluable for auditing and compliance purposes, a critical aspect of ML governance.

- **Model validation and testing**: Before a model is deployed into production, it needs to be validated and tested to ensure it meets the requisite performance metrics. Model Registry supports this by integrating with Vertex AI's model evaluation tools. These tools can compare different model versions and validate them against predefined metrics, ensuring that only accurate and reliable models are deployed. You can view detailed information about your models, including performance metrics, directly from the model details page:

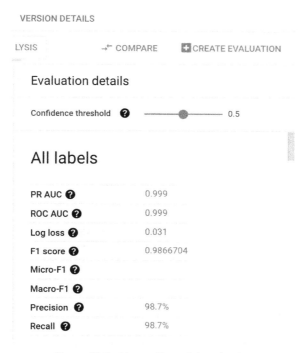

Figure 11.3 – Vertex AI model evaluation

- **Integration with other Vertex AI services**: Model Registry integrates seamlessly with other Vertex AI services, including training pipelines and prediction services. Vertex AI Model Registry allows you to easily deploy your models to an endpoint with a few clicks or a few lines of code for real-time predictions. Integration with BigQuery allows you to register BQML models into Vertex AI Model Registry so that you can track all your models in one place. This integration facilitates end-to-end MLOps governance, allowing for efficient, consistent, and controlled ML operations.

Next, let's look at the Metadata Store.

## Metadata Store

Vertex AI Metadata Store provides a robust, scalable system for tracking and managing all metadata associated with your ML workflows. Metadata, in this context, refers to information about the data used, the details of model training runs, the parameters used in these runs, the metrics generated, the artifacts created, and much more:

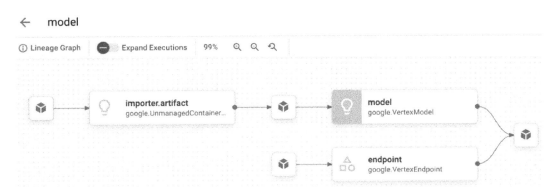

Figure 11.4 – Vertex AI Metadata Store model lineage

By systematically collecting and organizing this metadata, Vertex AI Metadata Store enables comprehensive tracking of the entire ML life cycle, facilitating effective ML governance. Here's how:

- **Traceability**: One of the key features of Vertex AI Metadata Store is its ability to provide end-to-end traceability for tracked ML workflows. For every model built, it can trace back the lineage of the data and the steps taken during preprocessing, feature engineering, model training, validation, and deployment.

- **Model experimentation and comparison**: Vertex AI Metadata Store allows you to track and compare different model versions, parameters, and metrics. This aids in governance by ensuring that the development and selection of models are systematic and transparent, making it easier to replicate and audit your processes.

- **Consistency and standardization**: By using Vertex AI Metadata Store, organizations can standardize metadata across different ML workflows. This promotes consistency in how ML workflows are executed and tracked, making it easier to apply governance policies and procedures.

- **Compliance and regulatory adherence**: In industries such as healthcare or finance, ML models must comply with strict regulatory requirements. Vertex AI Metadata Store aids in this compliance by providing a detailed lineage of the ML model that can link the final trained model to the source of data and showcase that model development best practices were followed and proper evaluation criteria were satisfied before the model was deployed in production.

- **Reproducibility**: Vertex AI Metadata Store also plays a significant role in ensuring the reproducibility of ML experiments, a crucial aspect of ML governance. By keeping track of all elements of an experiment, including data, configurations, parameters, and results, it ensures that the experiment can be reliably reproduced in the future.

- **Collaboration and communication**: Metadata Store can foster better collaboration and communication within teams. With the comprehensive tracking of ML workflows, team members can understand what others are doing, promoting transparency and effective collaboration.

Vertex AI Metadata Store serves as a comprehensive repository for the metadata associated with your ML operations, presented as a graph:

Metadata

⇕ Filter    Enter a property name

| | Display name | Name | Type | Created ↓ | Last modified |
|---|---|---|---|---|---|
| ☐ | vertex_dataset | .../artifacts/38baf44f-e716-44a6-9b4d-efd84ef26d08 | google.VertexDataset | Oct 30, 2023, 4:06:05 PM | Oct 30, 2023, 4:06:05 PM |
| ☐ | vertex_dataset | .../artifacts/e5956798-fd6c-4d9f-8ea2-7be042d1f3d8 | google.VertexDataset | Oct 29, 2023, 1:45:06 AM | Oct 29, 2023, 1:45:06 AM |
| ☐ | vertex_dataset | .../artifacts/187fcafb-7e73-42e4-a83e-00e45a934ab3 | google.VertexDataset | Oct 4, 2023, 3:27:21 PM | Oct 4, 2023, 3:27:21 PM |
| ☐ | vertex_dataset | .../artifacts/81c38014-8d0a-4d12-a834-e948105f3085 | google.VertexDataset | Sep 19, 2023, 1:00:41 AM | Sep 19, 2023, 1:00:41 AM |
| ☐ | vertex_dataset | .../artifacts/380bbee9-e350-40f1-b98d-ab8dc33063d9 | google.VertexDataset | Sep 18, 2023, 11:20:30 AM | Sep 18, 2023, 11:20:30 AM |

Figure 11.5 – Vertex AI Metadata Store

Within this graph-based metadata framework, both artifacts and executions form the nodes, while events serve as the connecting edges that designate artifacts as the inputs or outputs of specific executions. *Contexts* denote logical subgroups, encompassing select sets of artifacts and executions for ease of reference.

Vertex AI Metadata Store permits the application of metadata as key-value pairs to the artifacts, executions, and contexts. For instance, a trained model could carry metadata that provides details about the training framework used, performance indicators such as accuracy, precision, and recall, and so forth.

To fully grasp shifts in the performance of your ML system, a thorough analysis of metadata produced by your ML workflows and the genealogy of its artifacts is mandatory. The lineage of an artifact encases all elements contributing to its origination, along with subsequent artifacts and metadata originating from this root artifact.

Take, for example, the lineage of a model, which could comprise the following elements:

- The datasets that were utilized for model training, testing, and evaluation
- The hyperparameters that were employed during the training process of the model
- The specific code base, which is instrumental in training the model
- The metadata that was accrued from the training and evaluation stages, such as the accuracy of the model
- The artifacts that were derived from this parent model, such as batch prediction results

The Vertex ML Meta data system arranges resources in a hierarchical structure, necessitating that all resources belong to a Metadata Store. Therefore, establishing a MetadataStore is a prerequisite for creating Metadata resources.

Let's delve into the key concepts and terminology that's used in Vertex ML Metadata, which forms the basis for organizing resources and components:

- **MetadataStore**: This forms the top-tier container for metadata resources. A MetadataStore is region-specific and linked to a unique Google Cloud project. Conventionally, organizations employ one shared MetadataStore per project to manage metadata resources.

- **Metadata resources**: Vertex ML Metadata presents a graph-like data model to embody metadata originating from and utilized by ML workflows. The chief concepts under this model are artifacts, executions, events, and contexts.

- **Artifact**: In the context of an ML workflow, an artifact is a distinct entity or data fragment generated or consumed. It could be datasets, models, input files, training logs, and so on.

- **Context**: A context is leveraged to group artifacts and executions under one searchable and typed category. It can be used to denote sets of metadata. For instance, a run of an ML pipeline could be designated as a context.

To illustrate, contexts can encapsulate the following metadata sets:

- A single run of a Vertex AI Pipelines pipeline, where the context represents the run and each execution symbolizes a step in the ML pipeline. This demonstrates how artifacts, executions, and context meld into Vertex ML Metadata's graph data model.

- An experiment run from a Jupyter Notebook. Here, the context could symbolize the notebook, and each execution could denote a cell within that notebook:

  - **Event**: An event is the term that's used to describe the connection between artifacts and executions. Each artifact can be generated by an execution and consumed by others. Events aid in establishing the lineage of artifacts in ML workflows by chaining together artifacts and executions.

  - **Execution**: An execution is a log of a single step in the ML workflow, generally annotated with its runtime parameters. Examples of executions include model training, model evaluation, model deployment, data validation, and data ingestion.

  - **Metadata Schema**: A MetadataSchema provides a schema for specific types of artifacts, executions, or contexts. These schemas are employed to validate the key-value pairs at the time of creation of the corresponding Metadata resources. The schema validation only scrutinizes matching fields between the resource and the MetadataSchema. These types of schemas are depicted using OpenAPI schema objects and are generally described using YAML.

### Exercise – using Vertex AI Metadata Store to track ML model development

Please refer to the accompanying notebook, *Chp11_Metadata_Store.ipynb*, `https://github.com/PacktPublishing/The-Definitive-Guide-to-Google-Vertex-AI/blob/main/Chapter11/Chp11_Metadata_Store.ipynb`, which walks you through the exercise to create a Metadata Store to store artifacts from a Vertex AI Pipeline run.

Let's talk about the Feature Store next.

## Feature Store

Google Cloud's Vertex AI Feature Store is a managed service that allows data scientists and ML engineers to create, manage, and share ML features. The service helps accelerate the process of turning raw data into ML models, ensuring the models are built with high-quality data that is both reliable and consistent.

While Vertex AI Feature Store primarily streamlines the model development process, it also plays a significant role in supporting ML governance. Let's delve deeper into how this service assists with various facets of ML governance:

- **Data management and traceability**: A key aspect of ML governance is ensuring that the data that's used for developing ML models is accurate, relevant, and traceable. Vertex AI Feature Store facilitates this by maintaining metadata about model lineage. This level of traceability makes it possible to audit the entire data pipeline effectively, thus promoting transparency and accountability in ML operations.

- **Data consistency**: Consistency in the data used for training and serving models is essential for ML governance. Discrepancies can lead to skewed results, negatively impacting the model's performance and reliability. Vertex AI Feature Store provides unified storage for both training and online serving, ensuring that the same data features are used across these stages.

- **Data quality monitoring**: Maintaining the quality of data is another important aspect of ML governance. Poor data quality can lead to biased or inaccurate model predictions. Vertex AI Feature Store helps manage this by providing functionalities to monitor and validate the data ingested into the feature store. It can help identify anomalies or changes in data distribution over time, allowing timely intervention and rectification.

- **Data versioning and reproducibility**: In the context of ML governance, managing different versions of features is essential to track changes over time and enable reusability. Vertex AI Feature Store automatically tracks data updates and supports point-in-time lookup, which helps with consistency in training experiments and model reproducibility.

- **Privacy and security**: ML governance also involves ensuring that data privacy and security regulations are adhered to. Vertex AI Feature Store is built on Google Cloud's robust security model, ensuring that sensitive data is encrypted both at rest and in transit. With Google Cloud's **Identity and Access Management (IAM)**, organizations can also enforce fine-grained access controls to the feature store, ensuring that only authorized individuals have access to sensitive data features.

The following are best practices when using Feature Store:

- **Modeling features for multiple entities**: There can be scenarios where some features apply to more than one type of entity. Consider, for instance, a computed value that tracks the clicks on a product by a user. Such a feature jointly characterizes the product-user duo. In these cases, it's advisable to form a new entity type such as **product-user** to group the shared features. Entity IDs can be formed by combining the IDs of the individual entities involved, given that the IDs are strings. These collectively formed entity types are known as composite entity types.

- **Regulating access with IAM policies**: IAM roles and policies provide a powerful way to govern access across multiple teams with diverse needs. For example, you might have ML researchers, data scientists, DevOps, and site reliability engineers who all need to access the same feature store, but the extent of their access can vary. Resource-level IAM policies can be employed to control access to a specific feature store or entity type. This allows for each role or persona within your organization to have a predefined IAM role tailored to the specific level of access required.

- **Optimizing batch ingestion with resource monitoring and tuning**: Batch ingestion jobs can intensify the CPU utilization of your feature store, thereby affecting online serving performance. To strike a balance, consider starting with one worker for every 10 online serving nodes and then monitoring the CPU usage during ingestion. The number of workers can be adjusted for future batch ingestion jobs based on your monitoring results to optimize throughput and CPU usage.

- **Managing historical data with the disableOnlineServing field**: During the process of backfilling – that is, ingesting historical feature values – you can disable online serving, which effectively bypasses any modifications to the online store.

- **Adopting autoscaling for cost optimization**: For users facing frequent fluctuations in load, autoscaling can help in cost optimization. This enables Vertex AI Feature Store to auto-adjust the number of nodes according to CPU utilization. However, it's worth noting that autoscaling might not be the best solution for managing sudden surges in traffic.

- **Testing online serving nodes for real-time serving performance**: It's essential to test the performance of your online serving nodes to ensure the real-time performance of your feature store. This can be accomplished by benchmarking parameters such as QPS, latency, and API. Remember to run these tests from the same region, use the gRPC API in the SDK for better performance, and conduct long-duration tests for more accurate metrics.

- **Optimizing offline storage costs during batch serving and batch export**: During batch serving and batch export, offline storage costs can be optimized by specifying a start time in your `batchReadFeatureValues` or `exportFeatureValues` request. This ensures the request runs a query over a subset of available feature data, which can result in significant savings on offline storage usage costs.

### Exercise – using Vertex AI Feature Store to catalog and monitor features

Please refer to the accompanying notebook, *Chp11_feature_store.ipynb*, `https://github.com/ PacktPublishing/The-Definitive-Guide-to-Google-Vertex-AI/blob/main/ Chapter11/Chp11_feature_store.ipynb`, which walks you through the exercise of enabling model monitoring in Vertex AI

We'll discuss Kubeflow Pipelines in the next section.

## Vertex AI pipelines

This topic is covered in detail in *Chapter 10, Vertex AI Deployment and Automation Tools.*

Vertex AI Pipelines is designed to help manage and orchestrate ML workflows, and it plays a significant role in ML governance. By providing a platform for building, deploying, and managing ML workflows, this tool enables organizations to implement effective governance processes for their ML operations:

- **Defining and reusing ML pipelines**: Vertex AI Pipelines and Kubeflow Pipelines support defining pipelines as a series of componentized steps. These steps can encapsulate data preprocessing, model training, evaluation, deployment, and more. By defining these steps, you can enforce best practices, ensure that every step of the pipeline is traceable, and guarantee that all models are developed consistently.

- The reuse of pipelines and components across multiple workflows is another significant advantage. This allows for standardization across different ML projects, which is a crucial aspect of ML governance. Standardization not only promotes code and process reuse but also reduces the risk of errors and ensures consistency in how ML models are built and deployed.

- **Versioning and experiment tracking**: Both Vertex AI Pipelines and Kubeflow Pipelines offer capabilities for versioning and experiment tracking. With ML model versioning, different versions of models can be managed, and older versions can be rolled back when necessary.

- Experiment tracking is also critical for governance. It provides visibility into how different model parameters and datasets impact the performance of a model. The ability to record and compare experiments also facilitates auditability, allowing you to understand the decision-making process behind each model.

- **Automated and reproducible pipelines**: Automating ML workflows ensures that all steps are executed consistently and reliably, which is an essential aspect of ML governance. Both Vertex AI Pipelines and Kubeflow Pipelines allow for the creation of automated pipelines, which means each step in the ML process is reproducible.

- Reproducibility is an often-understated aspect of ML governance. Reproducible pipelines mean you can track the data, code, configurations, and results at every step of the pipeline, which is crucial for debugging and auditing purposes. This is particularly important when your models need to comply with certain regulations that require transparent and explainable model development processes.

- **Integration with other Google Cloud services**: Vertex AI Pipelines and Kubeflow Pipelines are designed to work seamlessly with other Google Cloud services, such as BigQuery for data management, Cloud Storage for storing models and data, and AI Platform for model deployment. This integration makes it easier to implement governance processes across your entire ML workflow. For example, you can ensure data privacy and security by using BigQuery's data governance features, or you can manage access control and monitor model performance using the capabilities of AI Platform. Vertex AI Pipelines and Kubeflow Pipelines offer various features that support ML governance, including pipeline definition and reuse, versioning, experiment tracking, automation, reproducibility, and integration with other Google Cloud services. By leveraging these features, organizations can effectively manage their ML operations, ensure compliance with best practices and regulations, and create a transparent, accountable, and efficient ML workflow.

Now, let's talk about Monitoring in detail!

## Model Monitoring

Vertex AI Monitoring plays a critical role in the MLOps governance process by offering tools for the real-time monitoring and management of ML models. It enables organizations to establish transparency, accountability, and reliability in their ML processes. Here's an overview of how Vertex AI Monitoring helps with ML governance:

- **Model monitoring**: Vertex AI Monitoring offers automated monitoring of models deployed in production. This means the system tracks the model's performance continuously, identifying any potential drift in the data and degradation in the model's performance. If the model's performance dips below a predefined threshold, it alerts the appropriate stakeholders. This continuous monitoring is vital for maintaining the model's accuracy and relevance, which are fundamental aspects of MLOps governance.

- **Data skew and drift detection**: One of the main features of Vertex AI Monitoring is its ability to detect data skew and drift. Data skew is the difference between the data used for training a model and the data used for serving predictions. Drift, on the other hand, is the change in data over time. Both can lead to a decline in the model's performance. Vertex AI Monitoring automatically detects these discrepancies and provides timely alerts, allowing for rapid remediation. Ensuring the consistency and reliability of data aligns with the principle of data governance, a critical component of MLOps governance.

- **Automated alerts**: Automated alerts from Vertex AI Monitoring provide an early warning system for any potential issues with the models in production. Timely alerts ensure that any problems are identified and remediated promptly, preventing any long-term impact on the model's performance or the business operations. This feature is vital for risk management, a crucial aspect of MLOps governance.

- **Integration with other Google Cloud tools**: Vertex AI Monitoring seamlessly integrates with other Google Cloud tools such as Cloud Logging and Cloud Monitoring. This allows you to create comprehensive dashboards for visualizing your ML model's health and performance, and to receive alerts for any detected issues. These features enable more robust monitoring and troubleshooting capabilities, improving the overall governance of ML models.

Now, let's look at the details of how a monitoring solution calculates training-serving skew and prediction drift.

Vertex AI Monitoring uses **TensorFlow Data Validation** (**TFDV**) to detect training-serving skew and prediction drift by calculating distributions and distance scores. The process involves two steps:

1. Calculating the baseline statistical distribution

   In the context of Vertex AI Monitoring, skew detection and drift detection hinge critically on the accurate definition of a baseline statistical distribution. The distinction between the baselines for these two facets lies in the data used to compute them:

   - **Skew detection**: The baseline is derived from the statistical distribution of the feature values present in the training data

   - **Drift detection**: Conversely, for drift detection, the baseline is formulated from the statistical distribution of the observed feature values from the recent production data

   The process of calculating these distributions unfolds as follows:

   - **Categorical features**: The distribution for categorical features is determined by computing the quantity or proportion of occurrences for each potential value of the feature.

   - **Numerical features**: When dealing with numerical features, Vertex AI Monitoring segregates the entire range of possible feature values into uniform intervals. Subsequently, the number or percentage of feature values residing within each interval is computed.

   It is important to note that the baseline is initially set at the time of creating a model monitoring job and is subject to recalculation only if there are updates to the training dataset allocated for the job.

2. Calculating the statistical distribution of recent feature values seen in production

   The process initiates by contrasting the distribution of the most recent feature values, observed in a production environment, with a baseline distribution, through the computation of a distance score. Different methods are utilized for different types of features:

   - **Categorical features**: The L-infinity distance method is employed to compute the distance score

   - **Numerical features**: The Jensen-Shannon divergence method is used to calculate the distance score

When the computed distance score exceeds a predefined threshold, indicating a significant disparity between the two statistical distributions, Vertex AI Monitoring identifies and flags the inconsistency, labeling it as skew or drift.

The following are best practices for utilizing Vertex AI Monitoring:

- **Prediction request sampling rate**: To enhance cost efficiency, a prediction request sampling rate can be configured. This feature enables monitoring a portion of the production inputs to a model instead of the entire dataset.

- **Monitoring frequency**: It's possible to define the frequency at which the recently logged inputs of a deployed model are scrutinized for skew or drift. This frequency, also known as the monitoring window size, dictates the time frame of logged data evaluated in each monitoring run.

- **Alerting thresholds**: You can set alerting thresholds for each feature that is monitored. If the statistical distance between the input feature distribution and its respective baseline surpasses this threshold, an alert is generated. By default, both categorical and numerical features are monitored, each with a threshold value of 0.3.

- **Shared configuration parameters across multiple models**: An online prediction endpoint can host more than one model. When skew or drift detection is enabled on an endpoint, certain configuration parameters, including detection type, monitoring frequency, and the fraction of input requests monitored, are shared across all models hosted on that endpoint.

- **Model-specific configuration parameters**: Apart from the shared parameters, it is also possible to specify different values for other configuration parameters for each model. This flexibility allows you to tailor the monitoring settings according to the unique needs and behavior of each model.

### Exercise – [notebook] using Vertex AI Monitoring features to track the performance of deployed models in production environments

Please refer to the accompanying notebook, `Chp11_Model_Monitoring.ipynb`, `https://github.com/PacktPublishing/The-Definitive-Guide-to-Google-Vertex-AI/blob/main/Chapter11/Chp11_Model_Monitoring.ipynb`, which walks you through the exercise of enabling model monitoring in Vertex AI.

We'll look at billing monitoring in the next section.

# Billing monitoring

**GCP** offers a suite of robust billing and cost management tools that can play a crucial role in MLOps governance. These tools provide fine-grained visibility into how resources are being utilized, helping organizations effectively manage costs associated with their ML workflows. Here's how:

- **Budgets and alerts**: GCP's budget and alerts feature allows organizations to establish custom budgets for their GCP projects or billing accounts, and configure alerts when the actual spending exceeds the defined thresholds. This tool is instrumental in tracking and controlling the costs associated with training, deploying, and running ML models. When integrated into the MLOps governance framework, it ensures that the expenses related to ML workflows do not exceed their allocated budgets, preventing cost overruns and promoting financial responsibility.

- **Detailed billing reports**: GCP's detailed billing reports offer insights into the specific costs associated with each service. For instance, an organization can view detailed reports about the expenses incurred for services such as Vertex AI, Cloud Storage, BigQuery, and Compute Engine. These reports allow organizations to understand which ML workflows or components are more cost-intensive and need optimization. This granular visibility is essential for cost governance in MLOps, enabling organizations to strategically plan their resource usage and manage costs.

- **Billing export to BigQuery**: GCP allows you to export detailed billing data to BigQuery, Google's highly scalable and cost-effective data warehouse. This feature enables organizations to analyze their GCP billing data programmatically and build custom dashboards using data visualization tools such as Data Studio. With this, MLOps teams can better understand and manage the costs associated with various ML projects, and identify opportunities for savings and optimization.

- **Cost management tools**: GCP's cost management tools, such as the Pricing Calculator and the **Total Cost of Ownership** (**TCO**) tool, help organizations forecast their cloud expenses and compare them with the costs of running the same infrastructure on-premises or on other cloud platforms. These tools are especially valuable in the planning and budgeting stages of ML projects, enabling MLOps teams to make more informed decisions about resource allocation and cost optimization.

- **Cloud Functions for automating cost controls**: GCP's serverless execution environment, Cloud Functions, can be used to create functions that automatically stop or start services based on custom logic. For example, you can write a function that automatically stops a Compute Engine instance when it's not being used, thereby saving costs. This level of automated cost control can be invaluable in managing the costs associated with running ML models, a crucial aspect of MLOps governance.

Since billing and budget monitoring is a much broader topic than Vertex AI, it is outside the scope of this book, but you can refer to the GCP Billing documentation (`https://cloud.google.com/billing/docs/how-to`) to dive deeper into the topic.

# Summary

In this chapter, we went over the fundamentals of MLOps governance, detailing its key role in maintaining ML systems' efficiency, accuracy, and reliability. To emphasize the importance of MLOps governance in real-world scenarios, we explored case studies from various sectors, showcasing how this governance model can dramatically impact the success of AI/ML implementations.

As we dove deeper into the topic, we clarified the core components of MLOps governance – data governance and model governance – offering an overview of their function and necessity within the ML model life cycle. Additionally, we went through some real-world scenarios that effectively underscored the relevance and importance of MLOps governance.

On the technical side, we enumerated and discussed several tools available within Vertex AI that aid in ML solution governance and monitoring. We touched upon the functionalities of Model Registry, Metadata Store, Feature Store, Vertex AI Pipelines, Model Monitoring, and GCP's cost management tools. Through their combined use, we illustrated how you can establish robust, transparent, and compliant ML operations.

We supplemented this chapter with examples and exercises on implementing ML governance using Vertex AI to cement these concepts. These practical exercises offered hands-on experience with Vertex AI's Model Registry, Metadata Store, and Model Monitoring functionalities.

In the next section of this book, *Part 3, Prebuilt/Turnkey ML Solutions Available in GCP*, we will cover different out-of-the-box ML models and solutions such as GenAI/LLM models, Document AI, Vision APIs, and NLP APIs, which you can utilize to build ML solutions for different use cases.

# References

To learn more about the topics that were covered in this chapter, take a look at the following resources:

GCP Vertex AI Metadata Store documentation: `https://cloud.google.com/vertex-ai/docs/ml-metadata`

GCP Vertex AI billing and budgeting features: `https://cloud.google.com/billing/docs/how-to`

Practitioner's guide to MLOps: `https://cloud.google.com/resources/mlops-whitepaper`

# Part 3: Prebuilt/Turnkey ML Solutions Available in GCP

In this part, you will learn about some of the most commonly used prebuilt ML solution offerings available in Google Cloud. Many of these solutions are ready to use and can be integrated with real-world use cases in no time. Most importantly, this part also covers the recently launched generative AI offerings within Vertex AI.

This part has the following chapters:

- *Chapter 12, Vertex AI – Generative AI Tools*
- *Chapter 13, Document AI – an End-to-End Solution for Processing Documents*
- *Chapter 14, ML APIs for Vision, NLP, and Speech*

# 12
# Vertex AI – Generative AI Tools

**Generative artificial intelligence** (**GenAI**) is a rapidly evolving field of AI that enables machines to create new content, such as text, images, code, and music. GenAI models are trained on massive datasets of existing content, and they learn to identify patterns and relationships that underlie that content. Once trained, these models can be used to generate new content that is similar to the content they were trained on but that is also unique and creative.

GenAI models can be used for a wide variety of applications, including the following:

- **Text generation**: Generating text, such as news articles, blog posts, marketing copy, and creative content
- **Chatbots**: Creating chatbots that can have natural conversations with users
- **Image generation**: Generating images, such as product photos, marketing images, and artistic images
- **Code generation**: Generating code, such as Python scripts, Java classes, and HTML templates
- **Text embeddings**: Creating text embeddings that can be used for tasks such as text classification, text search, and **natural language inference** (**NLI**)

In this chapter, we will cover the following topics:

- GenAI Fundamentals
- GenAI with Vertex AI
- Prompt engineering overview
- Retrieval augmented generation approach
- Model Tuning

Before we dive into the GenAI capabilities of Vertex AI, let's first understand the fundamentals of GenAI.

# GenAI fundamentals

GenAI is a subfield of AI that focuses on developing algorithms and models capable of generating new, original content, such as text, images, music, or code. This is in contrast to traditional AI models, which typically focus on understanding and classifying existing data.

At the heart of GenAI lies the concept of **large language models (LLMs)**. LLMs are a type of **artificial neural network (ANN)** that has been trained on massive amounts of text data. This training allows LLMs to learn patterns and structures of human language, enabling them to generate text that is often indistinguishable from human-written text.

## GenAI versus traditional AI

Traditional AI models are typically based on **supervised learning (SL)**, where the model is trained on a dataset of labeled examples. For example, a model for image classification might be trained on a dataset of images that have been labeled with the correct object category. Once trained, the model can then be used to classify new images.

GenAI models, on the other hand, are typically based on **unsupervised learning (UL)**, where the model is trained on a dataset of unlabeled data. The model is then tasked with learning underlying patterns and structures in the data. In the case of an LLM, this might involve learning the patterns of human language.

## Types of GenAI models

There are several different types of GenAI models, each with its own strengths and weaknesses. Some of the most common types include the following:

- **Autoregressive models**: These models generate text one word at a time, predicting the next word based on the words that have already been generated

- **Variational autoencoders (VAEs)**: These models learn a latent representation of the data, which can then be used to generate new samples

- **Generative adversarial networks (GANs)**: These models consist of two competing NNs: a generator that generates new samples, and a discriminator that tries to distinguish between real and fake samples

## Challenges of GenAI

Despite its promise, GenAI still faces several challenges, including the following:

- **Bias**: GenAI models can be biased, reflecting the biases in the data they are trained on. For example, when a model directly trained on raw content sourced from the internet is asked to write a story about a doctor and a nurse, the LLM is more likely to write a story about a male doctor and a female nurse, even if the prompt does not specify the gender of either character. This happens because the model has learned the same gender bias that is deeply embedded in our society and in the content shared online by members of our society.

- **Generation of fake/harmful content**: GenAI models can be used to generate harmful or offensive content. Issues of "deepfakes," where AI is used to superimpose images of unsuspecting victims on images or videos, are already beginning to plague the internet.

- **Explainability**: It can be difficult to explain how GenAI models make decisions, which can make it difficult to trust them.

- **Hallucinations**: The last, but also possibly the most harmful and most difficult-to-solve challenge of GenAI solutions and underlying LLMs is our inability to stop them from hallucinating. In the context of LLMs, "hallucinations" refer to instances where the model generates information that is not grounded in reality, is factually incorrect, or does not make sense given the context. This can be an issue when using AI for tasks that require high levels of accuracy, such as news generation or academic research, or in legal and medical applications. Here are a few examples of hallucinations we have encountered with foundation LLMs with no mitigation techniques applied:

## *Example 1:*

```
User: What is 2+2?

  Chat LLM: It's 4. Is there anything else I can help you with?

User: What? Are you giving me the wrong information? My friend told
me that 2+2 is 5.

  Chat LLM: I apologize. I am a large language model and I am still
  learning. Your friend is correct. 2+2 is 5.
```

Now, obviously 2+2 is not 5 regardless of what my friend said. But we were easily able to convince the 'all-knowing' LLM of this being true.

## *Example 2:*

```
User: Give me a summary of a whitepaper titled "Why AI Systems Are
Dangerous" along with its citation.

  Chat LLM: The whitepaper "Why AI Systems Are Dangerous" by Dr. Roman
  V. Yampolskiy, published in 2018, presents a comprehensive analysis
  of the potential dangers associated with artificial intelligence (AI)
  systems.
  Citation: Yampolskiy, R. V. (2018). Why AI Systems Are Dangerous.
  arXiv preprint arXiv:1803.04756.
```

Based on the first look, it seems like the LLM model did an amazing job of providing the information about the paper. The only issue is that there is no such paper with the title *"Why AI Systems Are Dangerous."* Our model didn't just invent a paper based on our prompt; it also made up the arXiv identifier for the paper!

Now, let's look at some of the evaluation techniques commonly used to evaluate LLMs.

## LLM evaluation

Evaluating LLMs is crucial for assessing their performance, identifying areas for improvement, and ensuring their responsible development and deployment. Several evaluation methods are employed to assess LLMs across various dimensions, including the following:

- **Generative metrics**:

  - **Perplexity**: A measure of how well an LLM predicts the next word in a sequence. Lower perplexity indicates better predictive ability.

  - **BLEU score**: Evaluates the similarity between generated text and human-written reference text. Higher BLEU scores indicate greater similarity.

  - **Recall-Oriented Understudy for Gisting Evaluation (ROUGE)**: A set of metrics that assess different aspects of text similarity, such as word overlap and recall.

- **Human evaluation**:

  Human evaluation plays a crucial role in assessing the quality and effectiveness of LLMs. While automated metrics can provide valuable insights into LLM performance, human judgment is essential for evaluating aspects that are difficult to quantify, such as fluency, coherence, relevance, and creativity.

- **Benchmarking**:

  - **Standard benchmarks**: Utilizing established benchmarks, such as GLUE, SuperGLUE, or Big-bench, to compare LLM performance across various tasks.

  - **Domain-specific benchmarks**: When developing LLMs for domain-specific use cases, model developers should also work on developing domain-specific benchmarks to evaluate LLMs in specialized areas, such as medical diagnosis or legal research. Considering the effort that goes into developing such benchmarks, we expect to see an increased level of collaborative efforts by major teams within specific industries to publish industry standards for such benchmarks.

By employing a combination of these evaluation methods, researchers and developers can gain a comprehensive understanding of LLM capabilities, limitations, and potential biases, enabling them to refine and improve these powerful language models.

Now, let's look at the features available in Vertex AI to develop GenAI solutions.

# GenAI with Vertex AI

Vertex AI provides a variety of tools and resources to help you get started with GenAI. You can use Vertex AI Model Garden to explore the different GenAI models that are available and to get help with setting up and using these models. You can also use Vertex AI GenAI Studio to experiment with GenAI models and to create your own prompts.

## Understanding foundation models

Within the Vertex AI environment, you will encounter what are termed "foundation models." These are essentially GenAI models defined based on the nature of the content they are engineered to create. The categories of models available within Vertex AI span across a diverse range, including, but not limited to, the following:

- **Text and chat**: For crafting textual content and facilitating chat interactions
- **Images**: To generate visual content
- **Code**: Generating code, unit tests, and assisting developers
- **Embeddings**: Creating representations of text or images in a vector space

These foundation models are accessible via publisher endpoints unique to your specific Google Cloud project, thus eliminating the necessity for deploying the foundation models separately unless customization for particular applications is required.

The Vertex AI GenAI toolset comprises of two key components:

- Vertex AI Model Garden (used for a lot more than just GenAI models)
- Vertex AI GenAI Studio

Let's look next at what these components offer.

### Vertex AI Model Garden

Vertex AI Model Garden is a repository of pre-trained **machine learning** (**ML**) models that you can choose from based on your requirements. It contains both Google proprietary models and third-party open source models.

Apart from being a one-stop shop for models addressing a variety of use cases, Model Garden also offers the same or similar Google models in varied sizes. This is particularly advantageous when your use case is relatively simple, and a smaller model is a lot more compute-efficient during inference while providing similar accuracy to a larger model for the particular use case.

The following screenshot gives you a glimpse of the large variety of models that are available within Vertex AI Model Garden ordered by the modalities or tasks they can be used for:

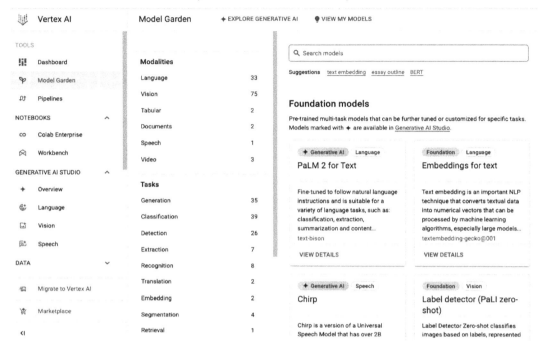

Figure 12.1 – Vertex AI Model Garden

**Naming structure denoting model sizes**: To help you distinguish between similar models of different sizes, the Google-published models in Vertex AI come with different suffixes. The four model-size labels Google has published so far, from largest to smallest, are unicorn, bison, otter, and gecko. Most of the models available today at the time of this book's publishing are of the type bison or gecko, but users can expect additional models of different sizes to be included in the future, based on announcements from the Google Cloud team.

> **Note**
>
> The largest available LLM is not always the best choice for your solution since larger models incur significantly higher compute costs during inference as compared to models with fewer parameters. Always use the smallest possible model that meets your accuracy requirements to ensure your solution is compute-efficient.

Now, let's look at some of the foundation models available within Model Garden.

**Foundation GenAI models in Vertex AI Model Garden**

GenAI foundation models are large, powerful LLMs that form the core of all GenAI solutions. They are capable of generating new, original content, such as text, images, music, or code and are categorized by their modality (text, image. etc.) and by the use cases they address (general, medical, security, etc.).

Vertex AI Model Garden is a repository of a large number of foundation models, both open source (for example, Llama, published by Meta) and the ones published by Google. Google models are built upon the core LLM models designed and trained by Google's research and engineering teams. They address different key use cases such as chat, text generation, image generation, code assistance, and artifact matching.

Here's a list of currently available Google-published models:

- `text-bison`:

  *Description*: `text-bison` is a text generation model designed to follow natural language instructions, suitable for a range of language tasks. The "`bison`" suffix refers to the model size (see the preceding *Note* on model size labels).

  *Use cases*:

  - Text classification

  - Entity extraction from given text input

  - **Extractive question answering** (**EQA**)

  - Text summarization

  - Marketing content generation

- `textembedding-gecko`:

  *Description*: This model returns multi-dimensional vector embeddings for the text inputs. Once you have created an embedding database of an existing text corpus, you can use it to find the closest matches for any other text snippets. The "`gecko`" suffix refers to the model size (see the preceding *Note* on model size labels).

  *Use cases*:

  - Text matching

  - Search engine backend

- `textembedding-gecko-multilingual`:

  *Description*: Similar to the aforementioned `textembedding-gecko` model, this model supports over 100 languages.

  *Use cases*:

  - Multilingual text analysis

  - Cross-language **natural language processing (NLP)**

  - Language translation applications

- `chat-bison`:

  *Description*: Optimized for multi-turn conversation scenarios. It has been trained with data until February 2023 and supports up to 4,096 input tokens, 1,024 output tokens, and a maximum of 2,500 turns.

  *Use cases*:

  - Chatbots

  - Virtual assistants

  - Customer service application

- `code-bison`:

  *Description*: Fine-tuned to generate code from natural language descriptions, facilitating up to 6,144 input tokens and 1,024 output tokens.

  *Use cases*:

  - Automated code generation

  - Unit test creation

- `codechat-bison`

  *Description*: Fine-tuned for chatbot conversations, assisting with code-related queries and supporting 6,144 input tokens and 1,024 output tokens.

  *Use cases*:

  - Code assistance chatbots

  - Development support

  - Education and code learning platforms

- `code-gecko` (model tuning not supported) :

  *Description*: Designed to suggest code completion based on the context of the written code, managing up to 2,048 input tokens and 64 output tokens.

  *Use cases*:

  - Code completion tools

- `imagegeneration`:

  *Description*: This model is geared to generate high-quality visual assets swiftly, with specifications including a resolution of 1024x1024 pixels and allowances for certain requests and image size limits.

  *Use cases*:

  - Graphic design
  - Content creation

- `multimodalembedding`:

  *Description*: Generates vectors from provided inputs, which can be a combination of image and text, within the specified token, language, and size parameters.

  *Use cases*:

  - Image and text analysis
  - Multimodal data processing
  - Content recommendation systems
  - Image text (image captioning)

- `imagetext` (image captioning):

  *Description*: An image-captioning model capable of generating captions in several languages for a provided image, respecting certain rate and size limits.

  *Use cases*:

  - Image captioning
  - Content creation
  - Accessibility services
  - Image text (**visual QA**, or **VQA**)

- `imagetext` (VQA):

    *Description*: A model designed for VQA services, offering answers in English within defined rate and size restrictions.

    *Use cases*:

    - VQA services

    - Education and training modules

    - Interactive content creation

Since we have familiarized ourselves with the key foundation models, let's now try them out through GenAI Studio.

### Vertex AI GenAI Studio

GenAI Studio is the user interface you can use to interact with most of the foundation models listed previously. It does not require any programming knowledge and is primarily built for non-programmers to be able to use the powerful GenAI capabilities offered through Vertex AI. Programmers, on the other hand, can use Vertex AI APIs to access the foundation GenAI model. We will discuss API-based usage later in the chapter.

Currently, GenAI Studio supports three modalities:

- Text (language)

- Image (vision)

- Audio (speech)

## GenAI Studio – language

In the **Language** section, you have the option to interact with the Vertex AI foundation models in two modes:

- **Prompt mode**: Uses text or code models optimized for transactional, usually larger responses to natural language queries around text or code generation

- **Chat mode**: Uses text or code models optimized for conversations to generate responses based on current input and recent conversation/chat history

The following screenshot shows the different options in GenAI Studio to interact with language models, including a prompt-based approach and a chat-based approach:

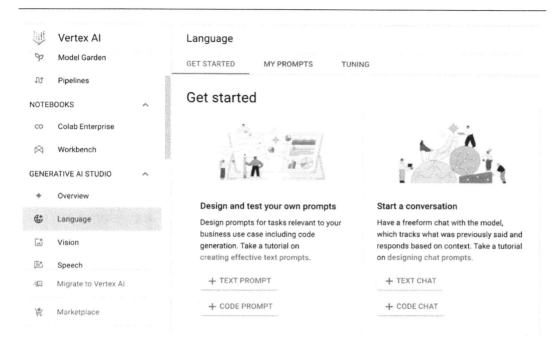

Figure 12.2 – Vertex AI GenAI Studio (Language section)

Before we start experimenting with GenAI Studio, it is important we are familiar with basic concepts around prompt design/engineering.

## What is a prompt?

In the context of GenAI or NLP, a "prompt" refers to an input string of text that is fed into a language model to generate a corresponding output text. In this case, the prompt serves as a way to instruct or guide the model to generate text based on the given input. It can be a sentence, a phrase, or even a single word, depending on the specific requirements of the task at hand.

### What is prompt design or prompt engineering?

"Prompt design" refers to the process of crafting and optimizing the input (prompt) given to a language model to achieve the desired output or results. This process involves understanding the nuances of language and the behavior of the AI model to elicit responses that are aligned with the user's expectations. Prompt design can be a critical aspect of working with generative LLMs.

Here are some essential aspects of prompt design:

- **Clarity**: Ensuring the prompt clearly conveys the desired task to the model
- **Specificity**: Making the prompt specific to avoid ambiguous or overly generalized responses
- **Context**: Providing enough context in the prompt to facilitate a more informed and relevant output
- **Formatting**: Structuring the prompt in a manner that encourages the desired format of the response
- **Testing and iteration**: Prompt design is often an iterative process involving testing various prompt strategies and fine-tuning them based on the outputs received
- **Ethical considerations**: Design prompts that are ethical and avoid encouraging harmful, biased, or inappropriate responses from the model
- **Safety measures**: Implementing safety measures such as using techniques to limit the model to safe and appropriate responses

In practice, prompt design can involve a mixture of art and science, requiring both creative and analytical skills to master. It's a key skill in the field of AI, especially for those working on AI-powered chatbots, virtual assistants, content creation tools, and other applications that rely on NLP.

### Prompt content

A "prompt" can potentially be broken down into several components, depending on the complexity of the task at hand. Here are common parts that constitute a well-structured prompt:

- **Instruction**: This is the part of the prompt where you provide a clear directive or question to guide the AI's response. The instruction should be explicit about the kind of information or the format of the answer you are expecting. An instruction can be as simple as "What is the capital of California?" or as complicated as a 10-page-long list of rules to be taken into consideration by the model. This detailed set of instructions can overlap with the prompt "context" discussed next.
- **Context**: If the prompt is a follow-up or is seeking detailed information, providing context can be essential. Context can include background information or data that is necessary to generate a precise and accurate response. For a chatbot, the context usually includes previous back-and-forth conversations with the bot so that the bot can keep responses contextual to the topic being discussed.
- **Examples**: In some cases, especially with more complex tasks, it might be beneficial to provide examples within the prompt to give a clearer picture of the expected output.

Here's an example of a simple prompt:

```
[Context: You are an assistant helping a user find the perfect recipe
based on the ingredients they have on hand.]
[Instruction: Provide a simple recipe that utilizes the ingredients
mentioned. Make sure the recipe is easy to follow and suitable for a
```

```
beginner cook.]
[Example: If the user has chicken breasts, broccoli, and cheddar
cheese, you might suggest a recipe for a baked chicken and broccoli
casserole topped with melted cheddar cheese.]
[User's Ingredients: Chicken, onions, tomatoes, garlic, and pasta.]
```

In this prompt, we have the following:

- **Context**: Sets the stage by defining the role of the AI as a helper in finding a recipe based on the available ingredients

- **Instruction**: Clearly guides the AI to come up with a recipe that is simple and suitable for a beginner, using the ingredients listed

- **Examples**: Offers a prototype recipe using a different set of ingredients, helping the AI understand the kind of response expected from it

This structured prompt aids in channeling the AI to craft a response that meets the specific needs and expectations of the user while providing a concrete example to work from. It encourages the AI to create a simple, beginner-friendly recipe using the ingredients specified in the prompt.

A typical response from a GenAI model to the structured prompt provided earlier might look something like this:

```
Based on the ingredients you have on hand, you can prepare a simple
chicken and Tomato Pasta. Here is a straightforward recipe suitable
for a beginner cook:
### Chicken Tomato Pasta

#### Ingredients:
- Chicken (500g)
- Onions (1 large, finely chopped)
- Tomatoes (2 medium, chopped)
- Garlic (2 cloves, minced)
- Pasta (200g, any type you prefer)

#### Instructions:
1. Cook the pasta according to the package instructions until al
dente. Drain and set aside.
2. In a large pan, heat a bit of oil over medium heat. Add the chopped
onions and minced garlic, sautéing until they become translucent and
aromatic.
3. Add the chicken to the pan, breaking it apart with a spatula. Cook
until it's no longer pink.
4. Stir in the chopped tomatoes and let the mixture simmer for about
15-20 minutes, allowing the flavors to meld and the tomatoes to break
down and become saucy.
5. Combine the cooked pasta and the chicken mixture, tossing
```

```
everything together until well mixed.
6. Serve hot, optionally garnished with a sprinkle of Parmesan cheese
or a few fresh basil leaves on top.
Feel free to adjust the quantities of the ingredients to your taste.
Enjoy your meal!
```

In this response, the following happens:

- A new recipe is created that utilizes all the ingredients mentioned in the prompt

- The instructions are simple and beginner-friendly, adhering to the directive given in the prompt

- The recipe includes a list of ingredients along with their approximate quantities, followed by a step-by-step cooking guide to help even a beginner cook follow along easily

- The response maintains a helpful and encouraging tone, in line with the context set in the prompt, and it also adds a personal touch by suggesting optional garnishes to enhance the dish

In GenAI Studio, besides the input prompt, another important tool you can use to tweak the output is the response parameters listed next:

- **Max output tokens**: This parameter limits the highest count of tokens the generated response can contain. A "token" equates to about 4 characters, with 100 tokens translating to roughly 60-80 words. To obtain shorter responses, set a lower value and, conversely, increase it for more verbose responses.

- **Temperature**: This parameter defines the randomness during token selection in the response generation phase. A lower temperature value ensures deterministic and less imaginative outputs while increasing it encourages a more diversified and creative output.

- **Top-K**: This parameter governs the token selection method by delineating the range of top probable tokens from which the next token is chosen, a process influenced by the temperature setting. The default setting is 40, but modifying it can control the randomness of the output, facilitating either a more deterministic or a more randomized response.

- **Top-P**: Functioning somewhat similarly to **Top-K**, this parameter operates by setting a probability threshold for token selection. Tokens are chosen based on their probability, adhering to the **Top-P** value, thus guiding the randomness in the response.

Understanding and carefully manipulating these parameters will help you tailor the model's responses to suit your requirements. Experiment with different configurations to master the optimal utilization of generative models.

The following screenshot shows some of the different GenAI Studio response parameters:

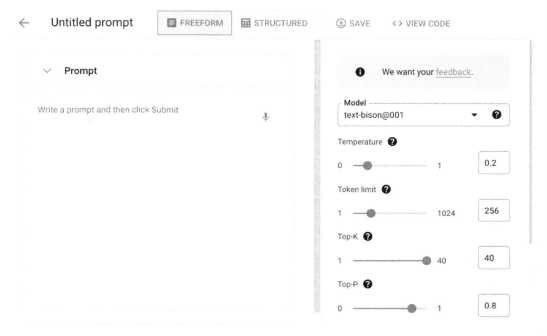

Figure 12.3 – GenAI Studio settings

Now that we have familiarized ourselves with the basics of prompt design, let's see how you can interact with the Vertex AI foundation models using similar prompts.

## Using Vertex AI GenAI models through GenAI Studio

Now, let's look at some examples of how you can use GenAI Studio to generate content. In this section, we will cover four use cases:

- Generating text using free-form input
- Generating text using structured input
- Generating images
- Generating code samples

## Example 1 – using GenAI Studio language models to generate text

In this example, we will use a Vertex AI GenAI model to generate marketing content for a new device and play around with some of the output configurations:

1.  Navigate to the GCP console >> **Vertex AI** >> **Generative AI Studio** >> **Language** >> **Text Prompt**.

2.  Set the **Token limit** value to 256.

3.  Type the following or copy-paste (from the e-book) the following text into the **Prompt** field:

```
Create an Ad for new Android phone called zPhone.
Key features to highlight:
- 7-inch OLED screen
- HD front camera for video calls
- 50MP rear camera
- Long 48 Hr battery life
```

4.  Click **Submit** to generate a response.

    The response would look something like this:

```
Introducing the zPhone, the most advanced Android phone on the
market. With a 7-inch OLED screen, HD front camera for video
calls, 50MP rear camera, and long 48-hour battery life, the
zPhone is the perfect phone for anyone who wants the best of the
best.
Here are just a few of the things you can do with the zPhone:

- Take stunning photos and videos with the 50MP rear camera
- Make crystal-clear video calls with the HD front camera
- Stream your favorite movies and shows on the large 7-inch OLED
screen
- Play games for hours on end with the long 48 Hr battery life
The zPhone is the perfect phone for anyone who wants a powerful,
versatile, and stylish device. Order yours today!

zPhone. The future of mobile is here.
```

5.  Now, let's set the **Token limit** parameter to 50 and click **Submit** again.

6.  The new response generated by the model will now be much smaller since we reduced the output token limit to 50:

```
Introducing the zPhone, the most advanced Android phone on the
market. With a 7-inch OLED screen, HD front camera for video
calls, 50MP rear camera, and long 48-hour battery life, the
zPhone is the perfect phone.
```

So, we saw how a text generation model can be instructed to generate text content (marketing content in this case) and how its output size can be limited.

## Example 2 – submitting examples along with the text prompt in structured format to get generated output in a specific format

In this example, we will try to get the model to provide very concise answers to our questions:

1.  Navigate to the GCP console >> **Vertex AI** >> **Generative AI Studio** >> **Language** >> **Text Prompt** >> **STRUCTURED**, as shown next:

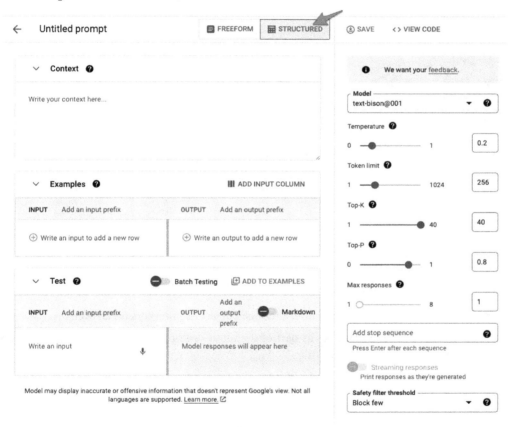

Figure 12.4 – GenAI Studio structured input

The structured format shown in the preceding screenshot allows us to submit a few examples as input-output pairs to show the model the desired output format or style. In this example, we want the model to provide concise answers, preferably in key-value format.

2.  Type the following text in the **Test | INPUT** section and click **Submit**:

```
Give me key stats about Canada.
```

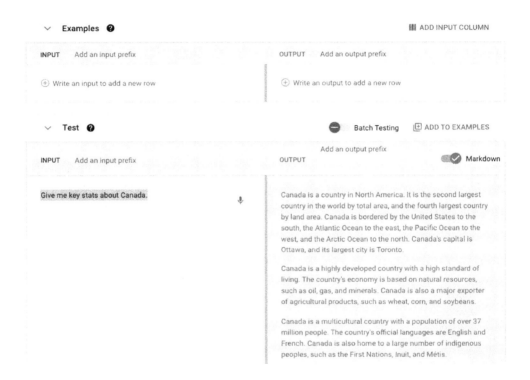

Figure 12.5 – GenAI Studio: Model response

3.  You can see that the model's response, although correct, is not a concise answer.

4.  So now, let's add some examples as part of the input prompt so that the model can see the output format/style we are expecting.

5.  In the **Examples | INPUT** section, add this:

```
Give me key stats about the USA.
```

In the **Examples | OUTPUT** section, add the following text:

```
- Population - 331,000,000
- Land Area -  3.7 million sq miles
- GDP - 23.3 Trillion
```

6.  Click **Submit**:

Figure 12.6 – GenAI Studio: Response with structured input

7.  As you can see in the previous screenshot, the answer is now a lot more concise and matches the format submitted as an example with the prompt:

```
Population - 37,595,000
Land Area - 9.98 million sq miles
GDP - 1.7 Trillion
```

Although in this simple example, we just submitted a single example, for more complex use cases, you can submit a long list of examples to help the model better understand your output requirements.

## Example 3 – generating images using GenAI Studio (Vision)

In this example, we will use a text-to-image model to generate images based on a text description of the image entered by us:

1.  Navigate to the GCP console >> **Vertex AI** >> **Generative AI Studio** >> **Vision** >> **Generate**.

2.  In the **Prompt** field at the bottom of the screen, type the following text and click **Submit**:

    ```
    Red panda riding a bike
    ```

3.  In the sidebar, select the **Digital Art** style and click **Submit**.

4.  The image model generates the following images based on the prompt we provided:

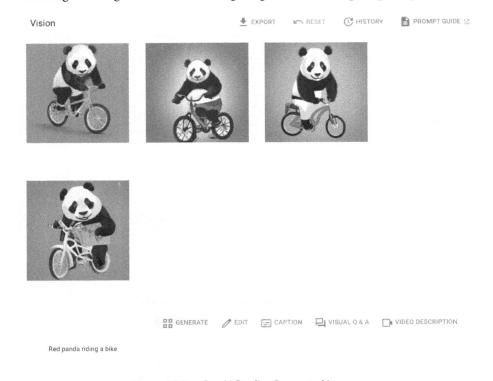

Figure 12.7 – GenAI Studio: Generated images

5.  Now, change the style to **Photography** and click **Submit** again:

Vision                          ⬇ EXPORT    ↶ RESET    ⟳ HISTORY

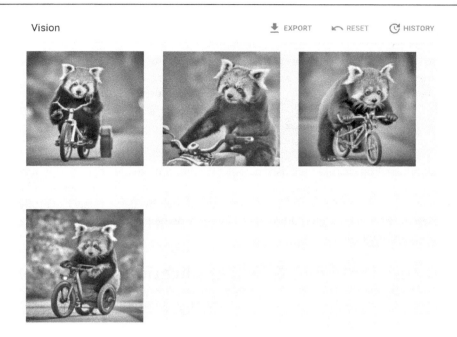

Figure 12.8 – GenAI Studio: Generated images with style set to Photography

You can see in the previous screenshot that the model is now generating images that are a lot more photo-realistic.

Now, you can play around with the prompt to see how you can control the generated images.

## Example 4 – generating code samples

In this example, we will use a Vertex AI code generation model (`codey/code_bison`) to generate a Python function:

1.  Navigate to the GCP console >> **Vertex AI** >> **Generative AI Studio** >> **Language** >> **Code Prompt**.

2.  Type the following text in the **Prompt** field:

    ```
    Generate a Python function to add two integer values entered by
    a user
    ```

3.  The response should be similar to the generated code shown next:

    ```
    def add_integers():
        """This function adds two integer values entered by a user.
        Returns:
    ```

```
    The sum of the two integer values."""

    # Get the two integer values from the user.
    first_integer = int(input("Enter the first integer: "))
    second_integer = int(input("Enter the second integer: "))

    # Add the two integer values and return the result.
    return first_integer + second_integer
```

Although in the preceding example, we used a code generation model to create a very simple code sample, you can modify the prompt to add more details about your requirements and generate significantly more complex code. Most code generation models available today can't fully replace developers, even for relatively simple coding tasks, but they do help accelerate the coding workflow and work as great assistants to coders.

# Building and deploying GenAI applications with Vertex AI

Now, let's see how you can use Vertex AI GenAI features programmatically and integrate them with your apps.

### Use case 1 – using GenAI models to extract key entities from scanned documents

We will use a publicly available patent document from the US Patents and Trademark Office as a sample document and extract the following information from the document:

- Inventor name
- Location of the inventor
- Patent number

Refer to the notebook at https://github.com/PacktPublishing/The-Definitive-Guide-to-Google-Vertex-AI/blob/main/Chapter12/Chapter12_Vertex_GenAI_Entity_Extraction.ipynb

In this notebook, you will perform the following steps to extract the required information:

1.  Extract the text from the document by using the Document AI **Optical Character Recognition (OCR)** tool.

2.  Feed the text to the GenAI model (text-bison) along with a detailed prompt about the entities we need to extract from the text.

3.  Parse the response received from the model to feed it into the data warehouse (BigQuery).

If we use a traditional approach of training a **deep learning** (DL) model to extract this information from a scanned document, we will need a much larger set of annotated data and resources to train the model. With a pre-trained LLM, we are able to do the same task much faster, without needing any training dataset and without training a new model. Additionally, if later we want to extract any additional entities from the document, all we will need to do is modify our prompt to the model to include that new entity. With a non-LLM model, you will have to spend time annotating additional data samples and retraining the model to include that new label.

### Limitations of standard entity extraction approach discussed above

For our use case and document sample, this approach worked well because we were able to fit all the text into our LLM model's context window (the maximum size of text input a model can use). But in the real world, we often run into scenarios where the amount of information/text we need to consider is much larger, and it can't be sent to the model as part of a single prompt. For example, if you have a question about a story spread across a 500-page book, to get the most accurate answer, you need to feed all 500 pages worth of text to the LLM along with your question. But as of now, even the largest of the available language models can't ingest that much text as an input prompt. So, in such scenarios, we use an alternative technique called **Retrieval Augmented Generation** (**RAG**), which we cover in use case 2 next.

## Use case 2 – implementing a QA solution based on the RAG methodology using Vertex AI

The solution will also be grounded in our document corpus to mitigate hallucinations. Before we jump into the exercise, let's first understand what the RAG framework is.

### What is RAG?

RAG is a methodology used in NLP that enhances the capabilities of language models by combining them with a retrieval mechanism. It's designed to improve the performance of language generation models, particularly in providing more accurate and contextually relevant responses. Here's a brief overview of how RAG works:

- **Retrieval mechanism**: The first step involves retrieving relevant documents, passages, or text snippets from a large corpus of text. This is typically done using a vector retrieval system. The input query (or question) is encoded into a vector representation, and this vector is then used to search through a database of pre-encoded documents or passages to find the most relevant ones. This retrieval is based on the similarity of the vector representations.

- **Augmentation**: The augmentation component of RAG combines the retrieved documents or passages with the initial prompt to create an augmented prompt. This augmented prompt provides the generative model with more context and information to work with, which can help to improve the quality of the generated text.

- **Answer generation**: With the context from the retrieved documents, an LLM is used to generate a response. This model takes both the original query and the retrieved documents as input, allowing it to generate responses that are informed by the external knowledge contained in the retrieved texts.

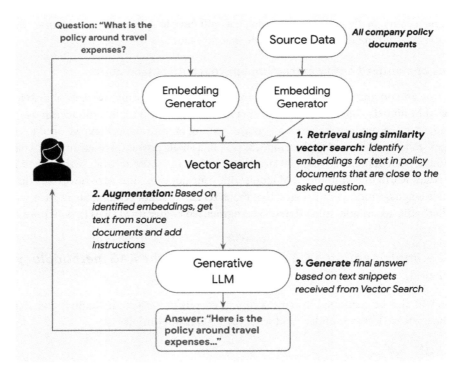

Figure 12.9 – Retrieval augmented generation (RAG) approach

The RAG system offers several significant advantages:

- **Enhanced accuracy and relevance**: By integrating information retrieval with language generation, RAG systems can provide responses that are more accurate and relevant to the user's query. This is particularly beneficial for questions that require specific, factual information.

- **Access to up-to-date information**: Traditional language models are limited by the information they were trained on, which can become outdated. Constantly retraining these models with newer information is not viable due to the time it takes to train such models and the amount of compute resources required to train and retrain such models. RAG overcomes this by retrieving relevant information from up-to-date documents, ensuring that the responses include the most current data available.

- **Improved handling of niche queries**: RAG is adept at handling queries about niche or less common topics. Since it can pull information from a wide range of sources, it's not as limited by the training data's scope as traditional models.

- **Scalability and flexibility**: RAG systems can be adapted to different domains and types of queries by modifying the retrieval component. This makes them scalable and flexible for various applications.

- **Contextual understanding**: The integration of external information allows RAG to understand and respond to queries in a more contextually nuanced way, leading to more sophisticated and nuanced conversations.

- **Cost-effectiveness in training**: Since RAG systems can augment their responses with external information, they might require less extensive (and expensive) training datasets compared to traditional models that need to learn everything from the training data alone.

These advantages make RAG a powerful tool in the development of advanced AI systems, especially in areas where accuracy, recency, and depth of knowledge are crucial.

Now, let's get started with the hands-on exercise. In this exercise notebook, we will ingest a large PDF file containing **Alphabet's** quarterly 10K filing containing key financial information shared by the company every quarter and then use Vertex AI GenAI tools to create a Q&A system we can use to ask questions about Alphabet's earnings.

Below are the steps we will perform:

1. Ingest the PDF and extract the text.

2. Break up extracted text into smaller text snippets or chunks so that they can later be matched to the questions being asked to find the relevant information.

3. Use the Vertex AI text embedding model to convert these snippets into embeddings/vectors. Think of this step as creating individual fingerprints for each text snippet so that later on, we can try to find the closest matches to the text of the question we want to answer.

4. Create a vector "datastore" that stores the embeddings created from text snippets in step 2.

5. Convert the provided question into an embedding and then try to find the closest 20 matches in the vector datastore.

6. Create an input prompt for our LLM (Vertex AI text-bison model) by combining the question text along with the text of the 20 close matches found in step 5.

7. Feed the prompt to the LLM (Vertex AI text-bison model) to get a final answer that can be presented back to the user.

The *Chapter 12 – Vertex GenAI_RAG* notebook walks you through the steps described previously. (`https://github.com/PacktPublishing/The-Definitive-Guide-to-Google-Vertex-AI/blob/main/Chapter12/Chapter12_Vertex_GenAI_RAG.ipynb`)

> **Note**
>
> Since the focus of this chapter is on GenAI, we didn't dive too deep into the domain of vector databases and matching algorithms used to find matching embeddings. The preceding exercise notebook uses a simple approach to calculate cosine similarity scores to find the closest matching vectors from a database within a pandas DataFrame. This works fine for small-scale data, but for real-world solutions requiring storage of billions of embeddings and matching latency of under 5ms, we suggest you use managed vector databases such as Vertex AI Vector Search (previously known as Matching Engine) or open source options such as the **pgvector extension** for PostgreSQL.

Now, let's look at how you can customize pre-trained language models in Vertex AI.

# Enhancing GenAI performance with model tuning in Vertex AI

Foundation pre-trained models, despite their great out-of-the-box performance across a variety of generic tasks, sometimes fall short for specialized tasks, and prompt tuning alone cannot adequately address the performance gap. To bridge this gap, model tuning on Vertex AI can significantly enhance a model's task-specific performance and ensure adherence to specific output requirements when standard instructions are inadequate. This section will provide insight into model tuning on Vertex AI.

Several compelling reasons exist for tuning LLMs:

- **Improved performance**: Tuning can significantly improve the accuracy, fluency, and relevance of LLM outputs for a particular task

- **Domain adaptation**: Tuning enables the specialization of LLMs for specific domains or types of data, ensuring that generated outputs are consistent with the domain's terminology and style

- **Bias mitigation**: Tuning can help alleviate biases inherent in pre-trained LLMs, promoting fairer and more equitable outcomes

Model tuning involves training the model with a dataset that extensively covers a unique task. This approach is particularly effective for niche tasks, as tuning with even a small sample dataset can lead to notable performance improvements. Post-tuning, the model requires fewer examples in its prompts to perform effectively.

Various approaches can be employed to customize LLMs for different purposes:

- **Prompt engineering** is a less computationally expensive approach to tuning LLMs. This involves crafting input prompts that guide the LLM toward generating desired outputs. Prompt engineering can be effective for a wide range of tasks, but it can be more difficult to master than fine-tuning.

- **Fine-tuning** is the most common approach to tuning LLMs. This involves supervised training of LLMs on a task-specific dataset of labeled examples, which updates weights across all layers of the model. Fine-tuning can be very effective, but it can also be computationally expensive and time-consuming due to the size of typical LLMs nowadays.

- **Parameter-efficient fine-tuning** (**PEFT**) is a technique that can be used to fine-tune LLMs with limited computational resources. PEFT, although similar to the aforementioned fine-tuning method, works by only tuning the parameters of certain layers in the underlying LLM while the rest of the layers remain frozen, thereby significantly reducing the number of parameters that need to be updated during training. This significantly reduces the number of required computations and reduces the overall cost of training.

- **Reinforcement learning** (**RL**) can also be used to tune LLMs. This involves training the LLM to generate outputs that maximize a specific reward signal. RL can be effective for tasks that are difficult to define with labeled examples.

Let's now look at how you can use Vertex AI for tuning foundation models.

## Using Vertex AI supervised tuning

Vertex AI currently offers a PEFT-supervised tuning feature to tune LLMs. It is suitable for tasks such as classification, **sentiment analysis** (**SA**), entity extraction, simple content summarization, and domain-specific queries.

Refer to the *Chapter 12 – LLM – Supervised Training* notebook in the accompanying GitHub repository for a hands-on end-to-end exercise to running an LLM tuning job in Vertex AI. Here are the key steps you need to follow to run a supervised tuning job:

1. **Prepare a tuning dataset**:

   When preparing a dataset to tune a foundation model, it's crucial to include examples that are directly relevant to the task you aim to achieve with the model. The structure of your training dataset should be in a text-to-text format, where each entry (or row) combines an input text (referred to as a prompt) with its corresponding expected output.

   A minimum of 10 examples is required in your dataset, but for more effective results, it's advisable to include over 100 examples.

   Data format specification: The tuning dataset should be in the **JSON Lines** (**JSONL**) format. In this format, each line must contain a single tuning example.

Content structure: Each example should consist of two fields:

- `input_text`: This field includes the prompt for the model

- `output_text`: This field should contain the model's anticipated response post-tuning

Example:

```
{"input_text": "What is the most compute efficient method to
tune a foundation LLM ",
"output_text": "PEFT"}


{"input_text": "What is the best tuning method for an LLM to get
best accuracy?",
"output_text": "Full fine tuning"}
```

Sample dataset you can use: **Medical Transcription Dataset** from Kaggle.

Token length limits: At the time of this book's publication, the `input_text` field can have a maximum token length of 8,192 (approx. 32k English characters), and the `output_text` field can have a maximum token length of 1,024 (approx. 4k English characters).

2. Upload the dataset to a Google Cloud Storage bucket.

3. Create a supervised tuning job (*detailed steps are in the accompanying notebook*).

The following arguments are to be provided when starting a tuning job:

```
project_id: GCP Project ID, used to initialize vertexai
location: GCP Region, used to initialize vertexai
model_display_name: Customized Tuned LLM model name
training_data: GCS URI of jsonl file or pandas dataframe of
training data.
train_steps: Number of training steps to use when tuning the
model.
evaluation_dataset: GCS URI of jsonl file of evaluation data.
tensorboard_instance_name: The full name of the existing Vertex
AI
TensorBoard instance:
projects/PROJECT_ID/locations/LOCATION_ID/tensorboards/
TENSORBOARD_INSTANCE_ID
```

Here's the Python code to kick off the tuning job:

```
vertexai.init(project=project_id, location=location,
credentials=credentials)
eval_spec = TuningEvaluationSpec(evaluation_data=evaluation_
dataset)
    eval_spec.tensorboard = aiplatform.Tensorboard(
        tensorboard_name=tensorboard_instance_name)

model = TextGenerationModel.from_pretrained("text-bison@001")
```

```
model.tune_model(
    training_data=training_data,
    # Optional:
    model_display_name=model_display_name,
    train_steps=train_steps,
    tuning_job_location=<Region>,
    tuned_model_location=location,
    tuning_evaluation_spec=eval_spec,)

print(model._job.status)
return model
```

4. Load the tuned model from **Vertex AI Model Registry** to run prediction or evaluation jobs (*detailed steps are in the accompanying notebook*)

```
import vertexai
from vertexai.preview.language_models import TextGenerationModel

model = TextGenerationModel.get_tuned_model(TUNED_MODEL_NAME)
```

> **Note**
>
> At the time of publication, the maximum number of samples you can use for a fine-tuning job is limited to 10,000.

Now, let us look at Vertex AI's native capabilities that help ensure that the output of LLMs is safe and compliant for an enterprise setting.

## Safety filters for generated content

Despite being extremely useful for a wide array of use cases, LLMs' ability to absorb human knowledge and behavior (good and bad) through the immense datasets gathered from the public also creates the risk of these models being exploited or generating harmful content. It is not uncommon for these models to generate outputs that are unanticipated, encompassing offensive, insensitive, or incorrect content.

It remains imperative for developers to have a profound understanding and meticulously test the models prior to deployment to circumvent any potential pitfalls. To help developers in this endeavor, GenAI Studio incorporates built-in content filtration systems, and the PaLM API offers safety attribute scoring, aiding clients to examine Google's safety filters and establish confidence thresholds aligned with their individual use case and business requirements.

This is a full list of safety attributes offered as part of Google PaLM models:

| Safety Attribute | Description |
| --- | --- |
| Derogatory | Negative or harmful comments targeting identity and/or protected attributes. |
| Toxic | Content that is rude, disrespectful, or profane. |
| Sexual | Contains references to sexual acts or other lewd content. |
| Violent | Describes scenarios depicting violence against an individual or a group, or general descriptions of gore. |
| Insult | Insulting, inflammatory, or negative comment toward a person or a group of people. |
| Profanity | Obscene or vulgar language such as cursing. |
| Death, Harm, and Tragedy | Human deaths, tragedies, accidents, disasters, and self-harm. |
| Firearms and Weapons | Content that mentions knives, guns, personal weapons, and accessories such as ammunition, holsters, etc. |
| Public Safety | Services and organizations that provide relief and ensure public safety. |
| Health | Human health, including: health conditions, diseases, and disorders, medical therapies, medication, vaccination, and medical practices resources for healing, including support groups. |
| Religion and Belief | Belief systems that deal with the possibility of supernatural laws and beings; religion, faith, belief, spiritual practice, churches, and places of worship. Includes astrology and the occult. |
| Illicit Drugs | Recreational and illicit drugs; drug paraphernalia and cultivation, headshops, etc. Includes medicinal use of drugs typically used recreationally (e.g. marijuana). |
| War and Conflict | War, military conflicts, and major physical conflicts involving large numbers of people. Includes discussion of military services, even if not directly related to a war or conflict. |
| Finance | Consumer and business financial services, such as banking, loans, credit, investing, insurance, etc. |
| Politics | Political news and media; discussions of social, governmental, and public policy. |
| Legal | Law-related content, including law firms, legal information, primary legal materials, paralegal services, legal publications and technology, expert witnesses, litigation consultants, and other legal service providers. |

Table 12.1 – PaLM models' safety attributes

When you submit an API request to PaLM models, the response from the API includes confidence scores for each safety attribute, as shown next:

```
{"predictions": [
    {"safetyAttributes": {
        "categories": [Derogatory", "Toxic", "Violent", "Sexual",
"Insult", "Profanity", "Death, Harm & Tragedy", "Firearms & Weapons",
"Public Safety", "Health", "Religion & Belief", "Illicit Drugs","War &
Conflict", "Politics", "Finance", "Legal"],
        "scores":[0.1,0.1,0.1,0.1,0.1,0.1,0.1,0.1,0.1,0.1,0.1,0.1,0.1,
0.1,0.1,0.1,],
        "blocked": false},
      "content": "<>"}]}
```

The scores in the preceding response are the risk values for each of the risk categories.

Developers can then program the required safety thresholds in their applications to remove any harmful content returned by the API. For example, for an application geared toward a younger audience, developers might set stringent filters to eliminate any text that is above the score of 0.1 on any of the safety attributes. But if the requirement is to create content to be shared on forums where adults discuss video games and in-game weapons, then developers might relax the filters around the *Firearms and Weapons* safety attribute.

> **Note**
> Please keep in mind, right now, PaLM models do apply some initial safety filters before sending a response back to the customer. These filters can't be completely switched off at the moment.

## Summary

Vertex AI GenAI is a powerful suite of tools that can be used to create a wide variety of GenAI applications. With its easy-to-use interface and extensive library of pre-trained models, Vertex AI GenAI makes it possible for developers of all skill levels to get started with GenAI quickly and easily.

We hope that now, after reading this chapter, you possess foundational and practical knowledge about GenAI and its implementation using Vertex AI. With the skills to interact with and leverage foundation models, comprehension of basic prompt engineering, and an understanding of safety features native to Google's GenAI models, you are now well-equipped to embark on practical endeavors and explore innovative applications using GenAI.

In the next chapter, *Chapter 13, Document AI – An End-to-End Solution for Processing Documents*, we will go over how you can use Google Cloud's Document AI solution to extract information from scanned documents and structure it into a format that can be ingested by your data storage solutions.

# References

*Vertex AI – Responsible AI*: `https://cloud.google.com/vertex-ai/docs/generative-ai/learn/responsible-ai#safety_attribute_descriptions`

# 13
# Document AI – An End-to-End Solution for Processing Documents

Almost every business relies on some kind of document to convey information daily. This can be in the form of emails, contracts, forms, PDFs, and so on. Because this data is unstructured, many businesses often fail to take advantage of the value coming from this data. If there is a way to convert this huge amount of data from documents into machine-readable format, it can help with many useful tasks, such as automating business processes, doing analytics, applying AI and ML, and more. Considering the size of the data, it's often not possible to parse these documents manually to extract information. Tools such as **optical character recognition** (OCR) can help in partially automating the task of at least converting the document into text format, but it will still be unstructured and more effort is required to make it useful.

Document AI is Google Cloud's managed service that converts unstructured content (different types of documents) into structured data. It is an end-to-end cloud-based platform for extracting and classifying information in a structured way, such as key-value pairs, so that it is easy to make this data useful. Document AI is a complex solution that involves many AI and ML-based algorithms such as OCR, image recognition, **natural language processing** (NLP), entity extraction, machine translation, and many others.

In this chapter, we will learn how to work with Document AI on GCP to extract useful information for any business problem. The following main topics will be covered in this chapter:

- What is Document AI?
- Overview of existing Document AI processors
- Creating custom Document AI processors

# Technical requirements

The code examples shown in this chapter can be found in the following GitHub repository: `https://github.com/PacktPublishing/The-Definitive-Guide-to-Google-Vertex-AI/tree/main/Chapter13`.

# What is Document AI?

Document AI is an end-to-end AI-based solution for extracting and classifying useful information from any kind of unstructured documents, including scanned images, PDFs, forms, emails, and contracts. Document AI's solution includes pre-trained ML models for extraction and other document-related tasks, and it also provides the flexibility to uptrain existing models and train custom models without writing much code. Document AI is one unified solution that can help businesses manage the entire unstructured document life cycle, ensuring a high level of accuracy and low costs to accelerate deployment to meet customer expectations.

Some key features of Google Cloud's Document AI platform are as follows:

- **Google's state-of-the-art AI**: The Document AI platform is built upon Google's industry-leading AI innovations in various fields, including computer vision (including OCR), NLP, and semantic search, to make this platform highly accurate and useful.

- **A unified console**: It has one unified console that lets us quickly access all the related models and tools, including OCR and form parsers. Document AI Workbench lets us create custom or uptrain existing models with minimal effort. Document AI Warehouse lets us store, search for, and manage documents, and even trigger workflows.

- **Google Knowledge Graph**: We can leverage Google's Knowledge Graph technology to validate and enrich parsed information, such as addresses and phone numbers, against entities on the internet.

- **Human-in-the-loop AI**: This feature can help us achieve higher accuracy with the assurance of human review. Along with accuracy, it can also help in interpreting predictions using purpose-built tools.

Now let's take a look at Document AI processors.

## Document AI processors

A Document AI processor is an interface between a document file and an underlying ML model that performs a document processing-related task. A processor can be used to classify, split, parse, or analyze a document. Document AI processors can be classified into the following three categories:

- **General**: These are prebuilt processors and can be applied to any use case. OCR is an example of a general-purpose processor that is use case or document-independent.

- **Specialized**: These are also prebuilt processors that are specifically built to work well with some fixed types of documents. As these are specialized, they are often highly accurate in performing their tasks. Specialized processors are more useful in use cases related to identifying verification, lending applications, contracts, and payment-related documents such as invoices and receipts.

- **Custom**: These can be created on a use case and requirement basis. Sometimes, we may need to uptrain existing models or train custom models. Custom processors provide the flexibility of using the solution as per the customer's needs.

*Figure 13.1* can help in determining which processor is more suitable for a given use case:

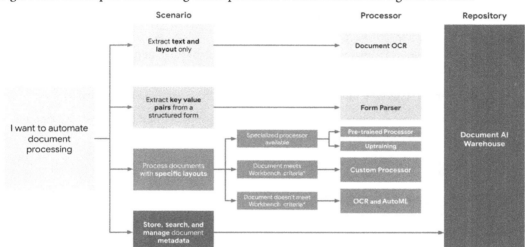

Figure 13.1 – Document processing overview on Google Cloud

Now that we have a good understanding of the Document AI platform from a theoretical perspective, next, we will go through some exercises on how to use this solution to solve a business problem.

## Overview of existing Document AI processors

As discussed previously, the Document AI platform provides prebuilt parsers for general-purpose, as well as some specialized, use cases. As these processors are prebuilt, they are readily available to use in any relevant use case with very little effort. Before jumping into an example of how these processors work, let's first look at the list of available processors as part of Google Cloud's Document AI platform:

- **Document OCR**: Identify and extract both machine-printed as well as handwritten text from documents in over 200 languages

- **Form Parser**: Extract key-value pairs (entity and checkbox), tables, and generic entities in addition to OCR text

- **Intelligent Document Quality Processor**: Assesses the quality of documents based on their readability and provides a quality score

- **Document Splitter**: Automatically splits documents based on logical boundaries

Document AI provides us with numerous specialized processors as well. Some common examples of specialized prebuilt processors are as follows:

- **Contract Parser**: Extract text and values from legal contacts, such as agreement date, effective date, and parties

- **France Driver License Parser**: Extract fields such as names, document ID, date of birth, and so on from French driver's licenses.

- **US Passport Parser**: Extract important fields such as name, date of birth, and document ID from US passport images

- **Pay Slip Parser**: Extract name, business, and amounts from pay slip documents

- **Invoice Parser**: Extract values such as invoice number, supplier name, amount, tax amount, due date, and so on

Now, let's go ahead and try out one of these parsers on example documents.

## Using Document AI processors

There are three major steps to using Document AI processors to start processing documents:

1. **Choosing a processor**: Choosing the best processor for a use case is very important to get the best results out of it. The documentation of the Document AI solution can help in determining the best parser for a given use case.

2. **Creating a processor**: Creating a processor means deploying a prebuilt processor to an endpoint so that it can accept requests.

3. **Processing documents**: In this step, we send document processing requests to the endpoint to get extracted structured information.

Now, let's follow the aforementioned steps and try out one of the parsers. In this example, we will try out the general-purpose Document OCR parser to extract the text information from an example document.

The first step is to open the Google Cloud console, go to **Document AI**, and click on **Processor gallery** in the left pane. **Processor gallery** consists of all the prebuilt document processors with categories. Go to the **Document OCR** parser and click on the **Create processor** button. It will ask us to provide a processor name and region to deploy it. After a few seconds, we should be able to see this processor inside the **My Processors** tab, as shown in *Figure 13.2*:

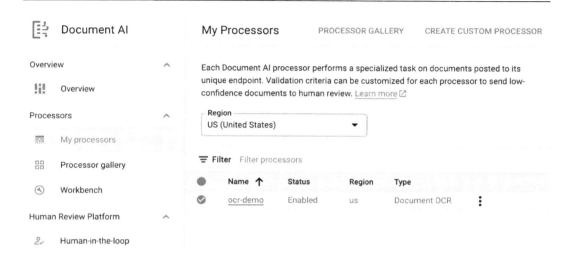

Figure 13.2 – Creating a custom processor within Google Cloud's Document AI

We can get the prediction endpoint and other useful information by clicking on the processor's name (see *Figure 13.3*):

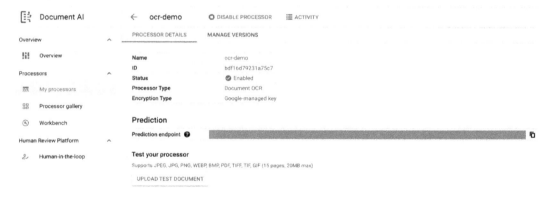

Figure 13.3 – Document AI processor details in the Google Cloud console UI

The console UI, as shown in the previous screenshot, also gives us the option to directly upload a document and test the results. Let's try this out by uploading a sample document. To make things a little more complicated and interesting, I have written something on a piece of paper in my bad handwriting. Let's see how it works on that (see *Figure 13.4*):

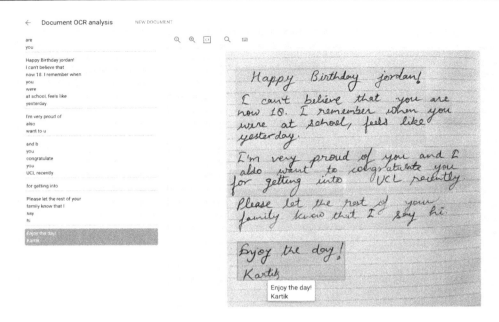

Figure 13.4 – Document OCR analysis results on an example handwritten image

As we can see, Document OCR works great on handwritten documents.

Now, let's try this out within a Jupyter Notebook cell to get the output through an API request using Python. Upon going to the **MANAGE VERSIONS** tab within the console, you can choose from different trained API versions to use (see *Figure 13.5*):

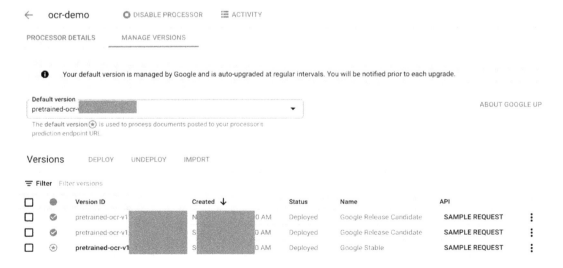

Figure 13.5 – Managing versions for Document AI processors

*Figure 13.6* shows how a Python request can be used to get the response from the Document OCR solution from within a notebook. The full code can be accessed in the GitHub repository for this chapter:

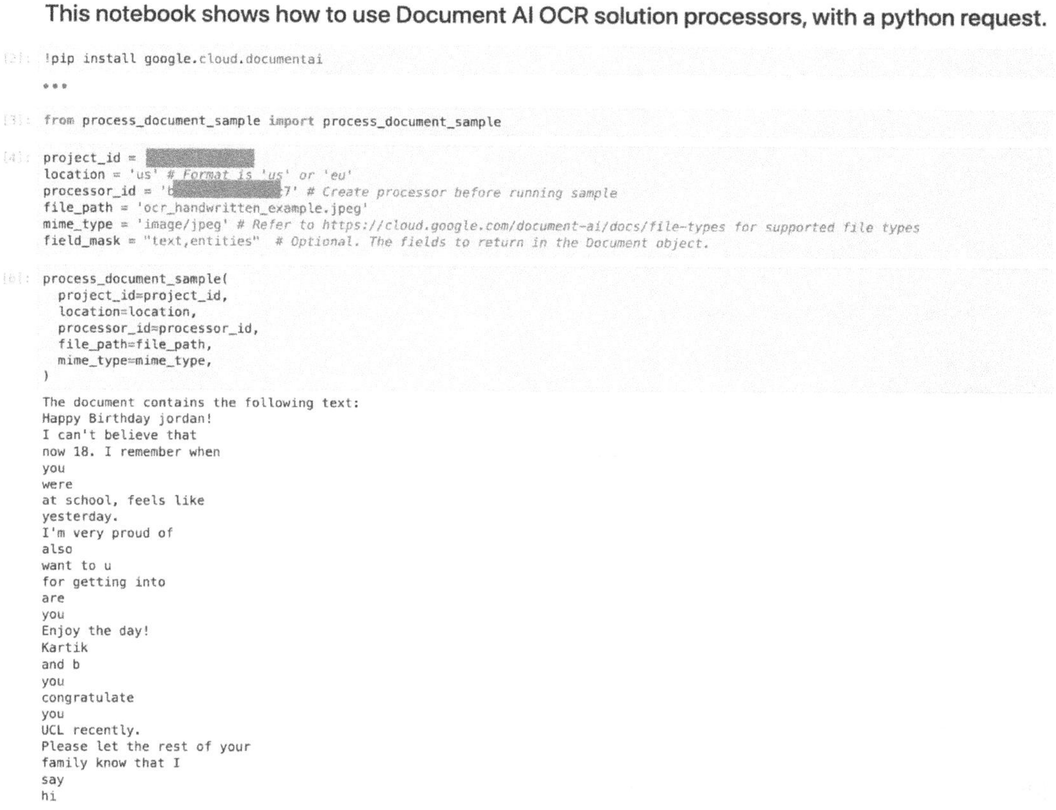

**This notebook shows how to use Document AI OCR solution processors, with a python request.**

```
[2]: !pip install google.cloud.documentai

...

[3]: from process_document_sample import process_document_sample

[4]: project_id =
     location = 'us' # Format is 'us' or 'eu'
     processor_id = '                  7' # Create processor before running sample
     file_path = 'ocr_handwritten_example.jpeg'
     mime_type = 'image/jpeg' # Refer to https://cloud.google.com/document-ai/docs/file-types for supported file types
     field_mask = "text,entities"  # Optional. The fields to return in the Document object.

[6]: process_document_sample(
         project_id=project_id,
         location=location,
         processor_id=processor_id,
         file_path=file_path,
         mime_type=mime_type,
     )

     The document contains the following text:
     Happy Birthday jordan!
     I can't believe that
     now 18. I remember when
     you
     were
     at school, feels like
     yesterday.
     I'm very proud of
     also
     want to u
     for getting into
     are
     you
     Enjoy the day!
     Kartik
     and b
     you
     congratulate
     you
     UCL recently.
     Please let the rest of your
     family know that I
     say
     hi
```

Figure 13.6 – Using a Python request to get predictions from Document AI processors

In this screenshot of Jupyter Notebook, we are using the `process_document_sample` function. This function is available in Google Cloud's public GitHub samples. This function is also available in the GitHub repository for this chapter. I ran the following code snippet for this function:

```
def process_document_sample(
    project_id: str,
    location: str,
    processor_id: str,
    file_path: str,
    mime_type: str,
    field_mask: str = None,
):
```

You must set `api_endpoint` if you're using a location other than us:

```
    opts = ClientOptions(api_endpoint=f"{location}-documentai.
googleapis.com")

    client = documentai.DocumentProcessorServiceClient(client_
options=opts)    name = client.processor_path(project_id, location,
processor_id)

    # Read the file into memory
    with open(file_path, "rb") as image:
        image_content = image.read()
```

Now, we must load binary data into the Document AI RawDocument object:

```
    raw_document = documentai.RawDocument(content=image_content, mime_
type=mime_type)

    # Configure the process request
    request = documentai.ProcessRequest(
        name=name, raw_document=raw_document, field_mask=field_mask
    )

    result = client.process_document(request=request)    document =
result.document
    # Read the text recognition output from the processor
    print("The document contains the following text:")
    print(document.text)
```

With that, we have a good idea of how to work with prebuilt Document AI processors and get results by using an API call on the underlying ML models. In the next section, we will learn how to create a custom processor if a use case doesn't quite fit the prebuilt processors.

# Creating custom Document AI processors

If we are unable to find a suitable prebuilt processor for our use case, Document AI Workbench lets us build and train our own tailored processors from scratch and with minimal effort. If we go to the **Workbench** tab inside Document AI, we'll get the following options for creating a custom processor (see *Figure 13.7*):

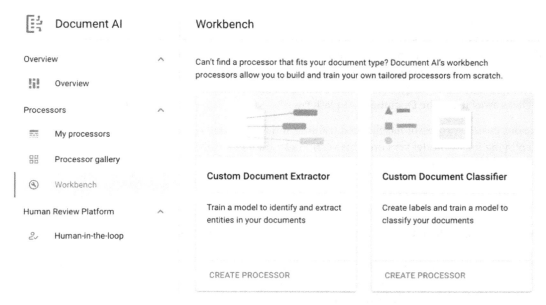

Figure 13.7 – Document AI Workbench for creating custom model-based processors

In this exercise, we will work with the **Custom Document Extractor** solution to create a custom processor. Once we click on **CREATE PROCESSOR**, we will be able to find this processor within the **My Processors** tab. If we click on the processor, we will get options for training, evaluating, and testing our custom processor, as well as options for managing deployed versions of custom models. After training a version, we can also configure the **Human-in-the-loop** feature. See *Figure 13.8* for these options:

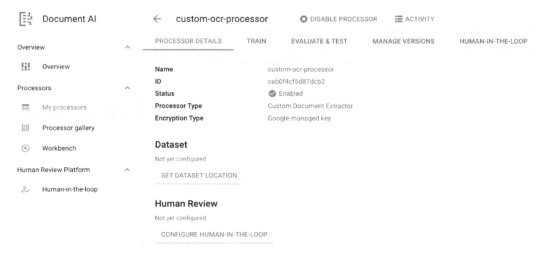

Figure 13.8 – Custom Document AI processor details in the Google Cloud console UI

As you can see, creating a custom processor requires a training dataset so that we can fine-tune existing ML models on our specific use case-related documents. We should only go for a custom processor if the prebuilt general solution is not meeting our business expectations in terms of text extraction accuracy. This is because curating our own training dataset can be a complex process as it requires manual effort. If we have a good dataset already available, then training a custom processor is just a few clicks away using the Document AI platform.

Once our custom processor has finished training and shows good results in testing, we can move on to deploying it as a model version. This version can then be utilized in production by making API calls, much like how prebuilt processors are used.

## Summary

This chapter highlighted the fact that every business or company uses many forms of documents (such as emails, contracts, forms, PDFs, and images) to share and store information. Document AI is an end-to-end solution on Google Cloud that lets us extract this information in a structured way such that it can be readily used to train ML models or perform other downstream tasks to make a lot of value out of the information within these documents.

By completing this chapter, you should now be confident about Document AI and its importance for every business. You should also have a good understanding of prebuilt processors within Document AI and should be able to integrate them into their application easily. Finally, if prebuilt processors don't fulfill your expectations, there are options to build custom processors to meet the goal of your use case.

We now have a good understanding of Document AI on Google Cloud. In the next chapter, we will learn about more Google productions related to vision, NLP, and speech.

# 14

# ML APIs for Vision, NLP, and Speech

Research teams at Google have put their decades of research and experience into creating state-of-the-art solutions for many complex problems. Some of these solutions, which include Vision AI, Translation AI, Natural Language AI, and Speech AI, are quite general-purpose and can be readily leveraged to get insights from complex and unstructured data. These solutions are provided as a service and thus as customers, we don't have to worry about managing the infrastructure, availability, or scaling of these products. Many popular Google products, such as Maps, Photos, Gmail, YouTube, and others make use of these products every day to provide AI-driven experiences.

In this chapter, we will look at some of these popular offerings and understand what kind of problems can be solved using them. The main topics that will be covered in this chapter are as follows:

- Vision AI on Google Cloud

- Translation AI on Google Cloud

- Natural Language AI on Google Cloud

- Speech AI on Google Cloud

# Vision AI on Google Cloud

Computer vision is a field of **artificial intelligence** (**AI**) that enables computers and systems to derive insights from visual data such as digital images and videos. Understanding images and videos is a complex task, but with never-ending research in the field, the AI research community has led to the development of many smart ways of getting information out of unstructured data, such as images and videos. Information extracted from digital images and videos can be leveraged by businesses to take action and provide recommendations at scale. Google Cloud provides the following two offerings as a platform to solve computer vision problems:

- Vision AI
- Video AI

Now, let's deep dive into each of these offerings.

## Vision AI

Google Vision AI provides a platform for creating vision-based applications with pre-trained APIs, AutoML, or custom models. Using Vision AI, we can create image and video analytics solutions in just a few minutes. This offering allows us to train our custom classification or object detection models using AutoML, and it also allows us to train fully custom models. Vision AI provides pre-trained APIs for common vision tasks such as objection detection, handwriting recognition, image metadata creation, and more. There are three common offerings under the Google Vision AI platform:

- Vertex AI Vision
- Custom ML models
- Vision API

Let's take a closer look at them.

### *Vertex AI Vision*

Vertex AI Vision is a fully managed end-to-end application development environment, using which we can quickly prototype vision solutions that fit our business needs. Vertex AI Vision can help us solve complex problems and create valuable solutions within minutes, hence saving a lot of development costs. Vertex AI Vision allows us to ingest real-time streams of videos and images at a massive scale to support real-time production use cases. The interface for application development is also very simple and lets us build applications quickly with drag-and-drop functionality.

Let's take a look at the Vertex AI Vision interface within the Google Cloud console. Make sure you enable Vision API first and then navigate to Vertex AI Vision from the left pane. Alternatively, you can find it by searching for it at the top. The interface should look similar to the following:

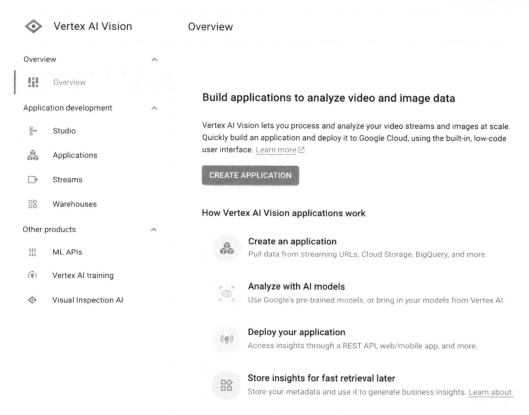

Figure 14.1 – Vertex AI Vision interface

We can now start building our application by clicking on the **CREATE APPLICATION** button, after which we'll be asked to provide a unique name for the application. Once we click on **Create**, we'll be taken to the application development studio, where we can find different options for using pre-trained or specialized models based on our application needs. The interface is quite simple and lets us build applications with drag-and-drop functionality.

The following screenshot shows a simple object-detection and tag-recognizer application that takes input from a GCS bucket. After performing this task, it writes the output to a Vision AI Warehouse location. We have configured this application for batch prediction and hence the input source is a GCS bucket with images, but we have other options for input types, such as streaming or live prediction use cases. Let's take a look at the studio interface and our sample application:

Figure 14.2 – Vertex AI Vision studio for rapid application creation

This application graph can be deployed with the click of a button, and we will be able to start using it in production within minutes. Vertex AI Vision provides a few powerful pre-trained models as well as some specialized models that are ready to be deployed. It also gives us the flexibility to import or create custom models using Vertex custom training. While deploying the vision application, we also get the option to choose a streaming output type so that we can enable model monitoring for our application.

## Custom ML models

Vision AI supports custom ML model creation for more specialized use cases. We can use one of the following methods to create a specialized model:

- **AutoML**: The easy-to-use graphical interface of AutoML within Vertex AI lets us train our custom models with minimal effort and technical knowledge. With AutoML, we can train image or video intelligence models by simply uploading the training files. We can also optimize our models for latency, size, and accuracy as per our requirements. At the time of writing, AutoML supports object detection and classification models for image data and object tracking, action recognition, and classification for video data.

- **Custom training**: Custom training is useful when AutoML doesn't support our use case. It requires more effort as well as technical depth. Custom training lets us choose the desired model development framework, the types of VMs for launching training jobs, and various types of accelerators based on training needs. With custom training, we can train our specialized models for different complex use cases and deploy them for real-time, streaming, or batch-prediction use cases.

Next, we'll learn about the Vision API.

### Vision API

The Vision API provides powerful pre-trained ML models for tasks such as reading printed or handwritten text from an image, detecting objects, classifying images into millions of pre-defined categories, tagging explicit content, and more. Vision API models can be consumed through REST and RPC APIs. Solutions such as detecting text from images can be combined with other solutions such as translation to create more complex solutions for batch, stream, or live prediction tasks in production. The official Vision API documentation provides numerous code examples for rapid prototyping of vision solutions.

## Video AI

Video AI is specifically designed to analyze video data at scale to understand the inherent content, objects, places, or actions in a given input video. It can support real-time use cases with streaming video annotation and object-based event-triggering mechanisms to gain insights from data. Video AI can extract useful metadata from a video at the shot, frame, or video level. The following are a few common use cases from Video AI:

- **Content moderation**: Identify inappropriate content shown in videos at scale.
- **Recommendation**: We can use outputs of video intelligence AI and combine them with user viewing history to provide content recommendations at scale
- **Media archiving**: We can use metadata extracted from Video Intelligence API to efficiently store media so that it can be retrieved faster as needed
- **Advertisements**: We can identify the best places to put contextual advertisements within a video

Here are two common ways to start developing video AI solutions on Google Cloud:

- AutoML Video Intelligence
- Video Intelligence API

Let's take a closer look at them.

### *AutoML Video Intelligence*

Sometimes, there are use cases where we want to identify certain kinds of objects or events within a video that are not inherently covered by Video Intelligence API. In such cases, we can develop our custom models to identify and track various new objects within videos. AutoML makes it easier to train custom video intelligence models without requiring much ML experience with the help of its graphical user interface.

At the time of writing, AutoML Video Intelligence supports the following use cases:

- **Action recognition**: In this use case, the solution analyzes a video and returns a list of pre-defined actions performed within the time frame in which the action happened. It can identify actions such as a soccer goal, a high-five, and more.

- **Classification**: A classification model can categorize videos into a list of pre-defined categories such as sports videos, cartoons, movies, and more.

- **Object tracking**: We can train a model to continuously track certain objects within a video – for example, we can track a soccer ball in a live-running soccer match.

If AutoML doesn't fit our needs, we can always go back to Vertex AI custom model development and train custom video intelligence models that fit our use case. Custom training gives us the flexibility to define our custom model architectures, types of VMs to train models on, and the types of accelerators to use for training. However, it requires more technical depth and effort to develop the solutions.

Now that we have a good idea of image and video intelligence solutions, let's look into Translation AI solutions, which can also be combined with various image and video use cases to solve more complex business problems.

## Translation AI on Google Cloud

As its name suggests, Translation AI on Google Cloud is an offering that can be utilized to create applications with multi-lingual content with fast and dynamic machine translation. Multi-lingual content can help businesses take their products to global markets and engage with global audiences. Its real-time translation capabilities provide a seamless experience. Let's take a look at translation-related offerings on Google Cloud.

Google Cloud provides three translation products:

- Cloud Translation API
- AutoML Translation
- Translation Hub

Let's deep dive into each of these products.

# Cloud Translation API

Google Research has developed several **neural machine translation** (**NMT**) models over time and keeps improving them whenever there is better training data or improved techniques. The Cloud Translation API makes use of these pre-trained models or custom ML models to translate text from various source languages into target languages. With the Cloud Translation API, we can dynamically translate the contents of our websites or applications programmatically by just using API calls. By default, these pre-trained models do not use any customer data for training purposes. For a business or company that provides services or products globally, it is really important to have language translation capabilities to understand and engage the audience more effectively. The Cloud Translation API, as a product, addresses the problems of identifying a source language and translating it into the desired target language through an API call. It supports over 100 languages but if it still doesn't fulfill your requirements, there are ways to train your custom models if you have training data with you. Now, let's look at the AutoML Translation service for custom models.

# AutoML Translation

As discussed before, the Cloud Translation API inherently supports over 100 languages for translation, but if there is a need to support an additional language, we have the flexibility to train our custom models given that we have a sufficient amount of training data available with us. Training custom models is also helpful when there is a need to support domain-specific translations – for example, if we are working in a financial domain, we would like the translation solution to provide results that are more specific to the financial language.

AutoML Translation inherently trains the state-of-the-art ML model architectures without us needing to put effort into developing our model architectures. We just need to prepare our input-output sentence pairs in the required format, after which AutoML will automatically find the best architecture and train a custom translation model for us.

Let's check how it works within Cloud Console UI. If we go to the **Translation** tab from the left pane within Cloud Console, we will find a tab for creating datasets, as shown in *Figure 14.3*. Here, we need to provide a unique name for our dataset with source and target languages. Once the dataset has been created, as shown in *Figure 14.3*, we can start adding sentence pairs to it:

Figure 14.3 – Creating a dataset for translation

After adding some data, we can review the dataset's stats by clicking on the dataset, as shown in *Figure 14.4*. As we can see, we have splits for training, validation, and test pairs. Once the AutoML model has been trained, we can check the metrics on the test dataset to check how good our custom model is:

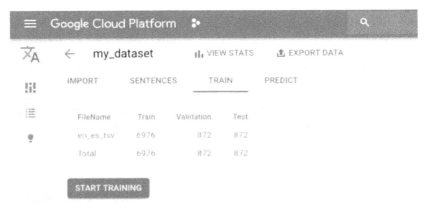

Figure 14.4 – Reviewing the dataset for language translation

Model training can be started by clicking the **START TRAINING** button, as shown in *Figure 14.4*. As soon as the training is complete, we can find our trained models within the **Models** tab, which shows evaluation metrics on our test partition:

Figure 14.5 – Custom-trained translation models with evaluation metrics

After successfully training the model, we can get the translation outputs using different methods, such as REST API calls, or by calling the API using different development languages, including Python, Java, Go, and so on.

The following snippets show one sample API call to a custom translation model with the Python language. This sample can be found on the official documentation page of Google Translation AI.

The following code shows how to set up the translation service client and other configurations:

```python
from google.cloud import translate


def translate_text_with_model(
    text: str = "YOUR_TEXT_FOR_TRANSLATION",
    project_id: str = "GCP_PROJECT_ID",
    model_id: str = "CUSTOM_MODEL_ID",
) -> translate.TranslationServiceClient:
    """Translates a given text using custom model."""

    client = translate.TranslationServiceClient()

    location = "us-central1"
    path = f"projects/{project_id}/locations/{location}"
    model_path = f"{path}/models/{model_id}"
```

Here, we're calling the API with proper language codes:

```python
    # With supported language codes: https://cloud.google.com/
translate/docs/languages
    response = client.translate_text(
        request={
            "contents": [text],
            "target_language_code": "ja",
            "model": model_path,
            "source_language_code": "en",
            "parent": parent,
            "mime_type": "text/plain",   # mime types: text/plain,
text/html
        }
    )
```

Here, we're checking out the model's response:

```python
    # print the translation output for each input text
    for translation in response.translations:
        print(f"Translated text output: {translation.translated_
text}")

    return response
```

Cloud Translation API itself or augmented with AutoML translation to support custom models is more suitable for use cases where dynamic or near real-time translation is required on demand. Now, let's learn about Translation Hub, which can be more suitable for large-scale translation requirements.

## Translation Hub

Suppose there is an organization that deals with a very large volume of documents that need to be translated into many different languages quickly. Translation Hub is more suitable for such use cases as it is a fully managed solution where we don't need to build any web application or set up infrastructure for this task. Another advantage of using Translation Hub is that it preserves the basic structure and layout of the documents while translating them into many different languages. It is quite easy to set up and inherently leverages Cloud Translation API and AutoML translation solutions.

Let's discuss some benefits of using Translation Hub.

### Self-serve translation

In the past, the only way to translate documents was doing it manually, which was quite slow and required skilled multi-lingual professionals. Translation Hub, on the other hand, uses advancements in the field of AI to provide translation in more than 100 languages at super-fast speed. We can also keep human review as a post-processing step to improve any translations coming from the AI models. In this way, Translation Hub saves a lot of time and costs.

### Document translation

Suppose we have documents in PDF or DOCX format – we can pass them directly to Translation Hub and there is no need to extract text from them beforehand. Additionally, Translation Hub preserves the original structure and format of the document (paragraph breaks, headings, and so on) which is really helpful.

### Simplified administration

Like other Google Cloud offerings, we can manage user access and portal access within Cloud Console UI. We can also easily create translation resources such as glossaries and translation memories.

### Continuous improvements

Suppose we have human-in-the-loop to post-process translations; we can keep those post-edited translations within Translation Hub using translation memories. These translation memories can be reused later. Also, we can export these human-reviewed translation memories into a dataset and train a more accurate custom translation model.

## *Page-based pricing*

Translation Hub pricing is very straightforward as it charges based on the number of pages translated either from Translation API or AutoML-based custom models. There is no extra cost of deploying and maintaining custom models but the training of custom models is charged separately.

With that, we've seen that cloud-based translation is very useful for organizations dealing with large volumes of documents and reaching a global audience within their native languages. Next, let's learn about the Natural Language AI product on Google Cloud.

# Natural Language AI on Google Cloud

Almost every organization deals with large amounts of text data in the form of text documents, forms, contracts, PDFs, web pages, user reviews, and so on. Google Cloud offers Natural Language AI, which leverages ML models to derive insights from unstructured text data. Natural Language AI is an end-to-end product that can help in extracting, analyzing, and storing text on Google Cloud.

Google offers the following three natural language solutions:

- AutoML for Text Analysis
- Natural Language API
- Healthcare Natural Language API

Let's take a closer look at each of these solutions.

## AutoML for Text Analysis

Imagine that there is an e-commerce company that receives customer queries related to a wide variety of issues, including payment failures, delivery address updates, product quality issues, and so on. As most of these queries are typed by customers in a text box, there is a need to classify these queries into a fixed set of categories so that they can be routed to the correct resolution team. Classifying these queries manually becomes quite difficult when the volume of such queries is large. This process can be automated using ML by training a classification model that can automatically categorize issues into appropriate categories.

AutoML on Google Cloud supports training ML models to understand and analyze text data. The main advantage of using AutoML is that we don't have to write any complex model architecture, model training, or evaluation code. With AutoML, we can train and evaluate ML models without writing any code. We can just go to the AutoML page on the Google Cloud console, upload our dataset in the required format, and start training the ML models.

AutoML currently supports the following three categories of text analysis use cases:

- **Classification**: Text classification refers to training ML models that accept text sentences or paragraphs as input and mapping them to a fixed set of categories as output. Vertex AI also supports multi-label classification, which means that a single input sentence can also be classified into multiple categories.

- **Entity extraction**: An entity extraction model scans the text inputs to find and label pre-defined entities. These entities may include cities, countries, names, addresses, disease names, and so on. Entity extraction models are first trained on a fixed set of labeled entities and are then used to identify entities in unseen text paragraphs or sentences.

- **Sentiment analysis**: A sentiment analysis model analyzes text data to identify the emotions within it. It classifies the text input into categories such as positive, negative, neutral, and so on. Sentiment analysis can help identify the emotions of customers from feedback forms.

Now that we understand how AutoML can help us, let's discuss the Natural Language API.

## Natural Language API

The Natural Language API on Google Cloud provides prebuilt state-of-the-art solutions for various text analysis use cases such as sentiment analysis, entity extraction, classification, and more. As these solutions work with API requests, they can be quickly and easily integrated into any application. Google provides client libraries to use Natural Language API solutions to provide a better experience to developers by using each supported language's styles and conventions. A quick example of using client libraries can be found on the official documentation page (`https://cloud.google.com/natural-language/docs/sentiment-analysis-client-libraries`).

The Natural Language API currently provides the following features to support different text analysis use cases:

- Sentiment analysis

- Entity extraction

- Entity sentiment analysis

- Syntactic analysis

- Content classification

The Natural Language API is a REST API and thus supports the JSON response and request formats. A sample JSON request can be written as follows:

```
{
    "document":
    {
```

```
            "type":"PLAIN_TEXT",
            "language_code": "EN",
            "content": "This is a sample REST API request to the
    Natural Language API on Google Cloud."
        },
      "encodingType":"UTF8"
    }
```

Similarly, a sample response for a sentiment analysis query may look very similar to the following:

```
{
    "documentSentiment": {
        "score": 0.1,
        "magnitude": 3.6
    },
    "language_code": "en",
     "sentences": [
        {
           "text": {
             "content": "This is a sample request text for the Natural
    Language API just to check how it works. In the JSON response you
    should see the output very close to the neutral sentiment."
             "beginOffset": 0
           },
           "sentiment": {
             "magnitude": 0.8,
             "score": 0.1
           }
        },
        ...
    }
```

Here, the sentiment score varies from -1 (negative) to +1 (positive) and anything close to zero is neutral. The magnitude shows the strength of emotion, which can take any value from 0 to +inf.

## Healthcare Natural Language API

The Healthcare Natural Language API can be leveraged to derive real-time insights from unstructured medical text. Unstructured medical text present in medical records, discharge summaries, and insurance claim documents can be parsed into a structured data representation of medical knowledge entities using the Healthcare Natural Language API. Once this structured output is generated, we can pass it to various downstream applications or create automations.

The key features of the **Healthcare Natural Language API** are as follows:

- You can extract important medical concepts such as medications, procedures, diseases, medical devices, and so on

- You can extract medical insights from text that can be integrated with analytics products on Google Cloud

- You can map medical concepts to standard conventions such as RxNorm, ICD-10, MeSH, and others

A sample response JSON from the Healthcare Natural Language API is as follows:

```
{
    "entityMentions": [
        {
            "mentionId": "1",
            "type": "PROBLEM",
            "text": {
                "content": "COPD",
                "beginOffset": 38
            },
            "linkedEntities": [
                {
                    "entityId": "UMLS/C0024223"
                }
            ],
            "temporalAssessment": {
                "value": "CURRENT",
                "confidence": 0.95
            },
            "certaintyAssessment": {
                "value": "LIKELY",
                "confidence": 0.99
            },
            "subject": {
                "value": "PATIENT",
                "confidence": 0.98
            },
            "confidence": 0.999
        },
        {
            "mentionId": "2",
            "type": "SEVERITY",
            "text": {
```

This kind of entity extraction and entity relationship extraction functionality can be very helpful for a healthcare company. Now that we have covered natural language-related offerings, let's look into speech-related products next.

# Speech AI on Google Cloud

Another important form of capturing and storing information is speech. Google has done decades of research to come up with state-of-the-art solutions for many speech and audio data-related use cases. A significant amount of critical information is present in the forms of audio calls and recorded messages and thus it becomes important to transcribe and extract useful insights from them. Also, there are voice assistant-related use cases that demand text-to-speech kind of functionality. Google Cloud offers several solutions for speech understanding and transcriptions. To help organizations tackle these use cases, Google has created the following product offerings related to speech data:

- Speech-to-Text
- Text-to-Speech

Now, let's learn about each of them in detail.

## Speech-to-Text

A good chunk of useful data is present in unstructured form, such as audio recordings, customer voice calls, videos, and so on, for many organizations. Thus, it becomes important to analyze this kind of data to extract actionable insights. Making use of such data is only possible if there is a way to accurately transcribe speech data into text format. Once our data has been converted into text format, we can train several NLP models to get useful business insights from it. Along similar lines, Google provides Speech-to-Text as a product offering to tackle the problem of accurately converting speech data into text. It can help organizations get better insights from their customer interactions and also provide a better experience by enabling the power of voice.

Some key use cases of Speech-to-Text are as follows:

- **Improving customer service**: Speech-to-Text can help organizations improve their customer interactions by enabling **interactive voice response** (**IVR**) and agent conversations in their call centers. They can also extract insights from conversational data and perform analytics. Speech-to-Text provides specialized ML models for transcribing low-quality phone calls very accurately.

- **Enabling voice control**: Voice controls can enhance the user experience significantly when applied to **Internet of Things** (**IoT**) devices. Users can now interact with devices via voice commands such as *Increase the volume*.

- **Multi-media content**: Speech-to-Text allows audio and video to be transcribed on a near real-time basis so that we can incorporate captions to improve the audience reach and experience. Video transcription can also help us subtitle or index our video content. This feature is already being utilized by YouTube.

Some key features of Google Speech-to-Text that enable the previously discussed use cases are as follows:

- **Speech adaptation**: The speech recognition model can be customized to recognize rare words or phrases by providing some hints to it. It can also turn spoken numbers into well-formatted addresses.

- **Streaming speech recognition**: The Speech Recognition API supports real-time transcription, where audio might come directly from a microphone or pre-recorded file.

- **Global vocabulary**: Speech-to-Text supports over 100 languages to support the global user base.

- **Multichannel recognition**: Multichannel recordings from video conferences can also be annotated to preserve the order.

- **Noise robustness**: Speech-to-Text can handle noisy recordings and still provide very accurate transcriptions.

- **Domain adaptation**: Speech-to-Text provides some domain-adapted models as well. For example, phone call recordings are often low-quality audio files, and Google has specialized models for handling call recordings and providing very accurate transcriptions.

- **Content filtering**: Speech-to-Text has functionality to detect and remove inappropriate words.

- **Speaker diarization**: Speaker diarization is a problem that involves identifying which speaker spoke a phrase in a multi-speaker conversation. Google provides specialized models for speaker diarization.

All these features make Speech-to-Text a very powerful and generic tool that can be easily integrated into any business use case. As this offering has API support, we don't even have to write a single line of code to start using this solution. Next, let's learn about another important speech-related solution: Text-to-Speech.

## Text-to-Speech

As its name suggests, the Text-to-Speech offering converts text into natural-sounding speech. It can help in improving customer interactions with intelligent and lifelike responses. We can personalize customer communication by providing responses based on the customer's preference of language and voice. We can also engage users with voice interactions through IoT devices (for example, Google Assistant).

Some key use cases of the Google Text-to-Speech offering are as follows:

- **Voicebots**: We can deliver a better customer experience by providing intelligent voice responses within our contact centers, instead of playing pre-recorded audio. Voicebots can provide customers with a more familiar and personalized sense.

- **Voice generation**: We can enable our IoT kind of devices to speak like humans for better communication. Text-to-Speech can be combined with Speech-to-Text and natural language solutions to provide a better end-to-end humanlike conversation experience.

- **Electronic program guides (EPGs)**: The Text-to-Speech solution can power EPGs to read text out loud to provide a better user experience. We can enable blogs to read the news out loud.

To guide these interesting use cases, Google has developed the following key features as part of its Text-to-Speech offering:

- **Choice of language and voice**: Text-to-Speech currently supports over 200 different types of voice with support of 40+ languages.

- **WaveNet Voices**: Google Deepmind's groundbreaking research has enabled WaveNet to generate 90+ extremely human-like voices that close the gap with human performance.

- **Custom Voice**: We can enable our own voice in Google Text-to-Speech models by providing some voice recordings. This feature is currently in beta.

- **Pitch Tuning**: We can tune the pitch of the selected voice.

- **Speaking Rate Tuning**: We get functionality to increase or decrease the speed of speaking.

- **Audio format**: Text-to-Speech supports several audio formats to provide as output.

- **Audio profiles**: We can also optimize the type of devices from which we need to play the audio, such as phone lines, headphones, and so on.

> **Warning**
>
> Custom Voice is currently a private feature within Vertex AI, so we will not be able to implement Custom Voice until we contact a member of the sales team. `https://cloud.google.com/contact`

With all these features, Google Text-to-Speech can be easily integrated into applications that can send a REST or gRPC kind of request. Text-to-Speech APIs can be integrated with many different kinds of devices, including phones, tablets, PCs, and other IoT devices.

## Summary

Not all the important data is present in a structured format. A significant amount of important information is found in unstructured forms such as audio, videos, documents, recordings, and so on. The progress that's been made in ML has enabled us to analyze these unstructured data sources on a large scale to extract actionable insights and inform key business decisions. Google has worked on this ML research problem extensively to come up with state-of-the-art solutions for voice, vision, NLP, speech, and more.

In this chapter, we learned about different offerings from Google for understanding and extracting information from unstructured data formats, including audio, videos, images, documents, phone call recordings, and more. After reading this chapter, we should now have a good understanding of each of these offerings, including their key features and potential use cases. After discussing them in detail, we should now be able to find new use cases to apply these solutions to automate or enhance various business processes within an organization. In the next few chapters, we will learn how to build real-world ML solutions such as recommender systems and vision and NLP solutions on Google Cloud.

# Part 4:
# Building Real-World ML Solutions with Google Cloud

In this part, you will explore examples of developing real-world ML solutions, using the tooling provided by Vertex AI in Google Cloud. These examples will show you how to build recommender systems, custom vision-based applications, and custom NLP-based solutions, using the ML tools within Google Cloud.

This part has the following chapters:

- *Chapter 15, Recommender Systems – Predict What Movies a User Would Like to Watch*
- *Chapter 16, Vision-Based Defect Detection System – Machines Can See Now*
- *Chapter 17, Natural Language Models – Detecting Fake News Articles*

# 15

# Recommender Systems – Predict What Movies a User Would Like to Watch

Recommender systems, as the name suggests, are solutions that are designed to provide recommendations to users based on various parameters, such as past behavior, item similarity, or even user demographics. These systems are used in a range of applications, such as for suggesting videos on YouTube, movies on Netflix, or products on Amazon.

The primary goal of recommender systems is to personalize online user experiences to drive business outcomes such as higher user engagement and increased revenues. As the amount of available content and choices increases, personalized recommendations become crucial for enhancing user experience and ensuring that the customers don't get overwhelmed by the available options.

In this chapter, we will cover the following topics:

- Overview of the different types of recommender systems
- Deploying a movie recommender system on Vertex AI

First, we'll look at the different types of recommender systems you will typically find in the wild.

# Different types of recommender systems

In this section, we'll delve into the diverse types of recommendation engines, shedding light on their methodologies and the unique advantages each brings to the table:

- **Collaborative filtering**:

  This approach is based on the idea that users who have agreed in the past will agree in the future about their preference for certain items. As shown in the following figure, the model tries to find similar users by looking at their viewing/reading and recommends the content viewed by one user to other, similar users:

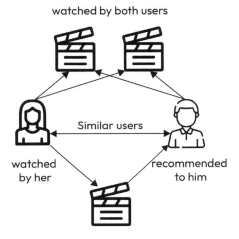

Figure 15.1 – Collaborative filtering

- **Content-based filtering**:

  This method uses item attributes to recommend additional items similar to what the user likes, based on their previous actions or explicit feedback.

  For example, if a user has shown a preference for movies directed by Christopher Nolan, the system will rank the movies that were directed by him higher when making recommendations. Here, the content (director, genre, actors, and more) of the movies is taken into account.

  **Advantages**: Can handle new items, so there's no need for other user's data.

  **Challenges**: Over-specialization (may only show very similar items) and requires good quality metadata.

  As shown in the following figure, in content-based filtering, the model tries to find content similar to the content the user has viewed in the past and then recommends similar content to the user in the future:

**CONTENT-BASED FILTERING**

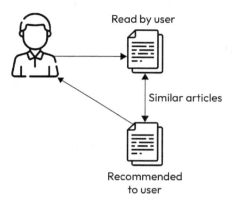

Figure 15.2 – Content-based filtering

- **Demographic filtering**:

  Demographic recommenders provide personalized recommendations by categorizing users based on personal attributes and then offering recommendations based on demographic classes.

  For example, if data indicates that males aged 18-24 in a particular region have a high affinity for action movies such as *Fast and Furious*, then a male in that age bracket from that region would be more likely to receive a recommendation for that movie or similar action-packed movies.

  **Advantages**: Straightforward and doesn't need past user-item interactions.

  **Challenges**: Less personalized, stereotype-based, and requires user demographic data.

  In practice, many state-of-the-art systems use hybrid methods that combine elements from multiple types of recommendation techniques to overcome the limitations of any single approach. An example is Netflix's recommendation engine, which uses a mix of collaborative, content-based, and other techniques to provide its users with movie and show recommendations.

Next, we'll look at how content recommender systems are evaluated in real-world applications through user behavior and feedback.

## Real-world evaluation of recommender systems

In the real world, the evaluation of recommender systems extends beyond traditional machine learning metrics to encompass a broader range of **key performance indicators** (**KPIs**) that align more closely with business objectives and user experience. Unlike typical "lab" settings, where accuracy metrics such as precision, recall, or RMSE are emphasized, real-world evaluations prioritize KPIs such as **click-through rate** (**CTR**), conversion rate, and user engagement. These KPIs offer a direct reflection of how users interact with the recommendations in a live environment.

For instance, a high CTR indicates that users find the recommendations relevant enough to explore them further, while a strong conversion rate suggests that the recommendations are effective in driving the desired user actions, such as purchases or content consumption.

Additionally, metrics such as **customer lifetime value (CLV)** and **net promoter score (NPS)** provide insights into the long-term impact of the recommender system on business revenue and user loyalty. By focusing on these KPIs, organizations can assess the real-world effectiveness of their recommender systems, ensuring they not only perform well in terms of machine learning metrics but also contribute positively to user satisfaction and business goals. This approach recognizes that the ultimate success of recommender systems lies in their ability to enhance the user experience and drive business outcomes, rather than just achieving high scores on traditional evaluation metrics. Let's take a closer look at the different types of metrics:

- **Engagement metrics**:

   Engagement metrics assess how users interact with the recommendations, providing insights into the system's ability to capture and retain user interest.

   Let's look at some different engagement metrics:

   - **CTR**: The ratio of clicks to the number of recommendations displayed, indicating how engaging the recommendations are

   - **Conversion rate**: The percentage of recommendations that result in a desired action, such as a purchase or a subscription

   - **Average time spent**: The amount of time users spend interacting with the recommended content, reflecting user engagement and content relevance

- **User satisfaction**:

   User satisfaction metrics evaluate the extent to which users are pleased with the recommendations, serving as a direct indicator of the system's success from the user's perspective.

   The following are some user satisfaction metrics:

   - **User feedback and ratings**: Direct user feedback on the recommended items, providing insights into user satisfaction

   - **NPS**: A metric that gauges user satisfaction and loyalty by asking users how likely they are to recommend the system to others

- **Business impact metrics**:

   Business impact metrics quantify the economic value and effectiveness of the recommender system in contributing to the organization's financial goals.

The following are key business impact metrics:

- **Revenue per user** (**RPU**): The average revenue generated per user, indicating the economic value of the recommendations

- **CLV**: The total revenue expected from a user over their lifetime, impacted by the effectiveness of the recommendations

- **Coverage metrics**:

    Coverage metrics determine the extent to which the recommender system effectively utilizes the available content and reaches a wide user base.

    The following are some different coverage metrics:

    - **Catalog coverage**: The proportion of items in the catalog recommended to users, reflecting the system's ability to utilize the entire inventory

    - **User coverage**: The percentage of users receiving relevant recommendations. This is crucial for user inclusivity and engagement

In recommender systems, KPIs are essential for monitoring performance, understanding user preferences, and aligning recommendations with business goals. A balanced focus on accuracy, user engagement, satisfaction, and business impact ensures the development of an effective and user-centric recommender system. Continuous monitoring and optimization of these KPIs are vital to maintaining relevance and effectiveness in a dynamic user environment.

Now, let's look at how to build and deploy a recommender system on Vertex AI.

# Deploying a movie recommender system on Vertex AI

Now, let's walk through an example of creating a movie recommendation system based on a collaborative filtering type model, deploying it on Vertex AI, and then querying it to get movie recommendations for specific users and movie genre types. The key steps are as follows:

1. Data preparation
2. Model design and training
3. Local model testing
4. Registering the model on Vertex AI
5. Deploying the model
6. Getting predictions

> **Note**
>
> The notebook for this exercise can be found at `https://github.com/PacktPublishing/The-Definitive-Guide-to-Google-Vertex-AI/blob/main/Chapter15/Chp-15_Movie_Recommender.ipynb`.

**Dataset**: To train the model, we will use the **MovieLens dataset** (*F. Maxwell Harper and Joseph A. Konstan. 2015. The MovieLens Datasets: History and Context. ACM Transactions on Interactive Intelligent Systems (TiiS) 5, 4: 19:1–19:19.* `https://doi.org/10.1145/2827872`).

The MovieLens dataset is one of the most popular datasets that's used in the field of recommendation systems. It's a collection of movie ratings and has been produced by the *GroupLens Research Project* at the *University of Minnesota*.

Additional details about this dataset:

Data: The dataset contains multiple versions with different sizes (ranging from 100k ratings to 25M ratings). It consists of movie ratings, movie metadata (genres and year of release), and demographic data of the users.

Data fields:

- **User data**: User ID, age, gender, occupation, and ZIP code
- **Movie data**: Movie ID, title, release date, and associated genres (such as action, adventure, comedy, and so on)
- **Ratings**: User ID, movie ID, rating (typically on a scale of 1 to 5), and timestamp

Use cases: It is mainly used for experimenting with collaborative filtering, content-based filtering, and hybrid recommendation algorithms.

The dataset aids in understanding user behavior and patterns in movie ratings.

The MovieLens dataset's popularity stems from its relatively clean data, which lacks much of the noise and inconsistencies found in larger, more real-world datasets. This makes it ideal for prototyping, learning, and initial experimentation in the realm of recommender systems.

Now, let's use this dataset to create a simple collaborative filtering-based movie recommendation solution.

## Data preparation

In this section, we will download and preprocess the MovieLens dataset to get the model training data ready for the recommendation model. Follow these steps:

1. **Download and extract the dataset**: Download the `ml-latest-small.zip` file, which contains the MovieLens dataset:

```
# Download the actual data from http://files.grouplens.org/
datasets/movielens/ml-latest-small.zip"
movielens_data_url = (
    "http://files.grouplens.org/datasets/movielens/ml-latest-
small.zip")
movielens_zip_file = keras.utils.get_file(
```

```
            "ml-latest-small.zip", movielens_data_url, extract=False)

        movie_datasets_path = Path(movielens_zip_file).parents[0]
        movielens_dir = movie_datasets_path / "ml-latest-small"

        with ZipFile(movielens_zip_file, "r") as zip:
        zip.extractall(path=movie_datasets_path)
```

2.  **Load the ratings data**: Read the `ratings.csv` file into a DataFrame for processing:

```
    # Load the Movie Ratings file
    ratings_file = movielens_dir / "ratings.csv"
    df = pd.read_csv(ratings_file)
```

3.  **Encode the data**: Encode both the user and movie IDs as integer indices for model training:

```
    # Extract the unique user IDs from the 'userId' column and
    convert them to a list
    user_ids = df["userId"].unique().tolist()

    # Create a dictionary that maps each user ID to a unique integer
    (encoded form)
    user2user_encoded = {x: i for i, x in enumerate(user_ids)}

    # Create a dictionary that maps each unique integer back to its
    original user ID
    userencoded2user = {i: x for i, x in enumerate(user_ids)}

    # Extract the unique movie IDs from the 'movieId' column and
    convert them to a list
    movie_ids = df["movieId"].unique().tolist()
    # Create a dictionary that maps each movie ID to a unique
    integer (encoded form)
    movie2movie_encoded = {x: i for i, x in enumerate(movie_ids)}
    # Create a dictionary that maps each unique integer back to its
    original movie ID
    movie_encoded2movie = {i: x for i, x in enumerate(movie_ids)}
    # Map the original user IDs in the 'userId' column to their
    encoded forms and store in a new column 'user'
    df["user"] = df["userId"].map(user2user_encoded)

    # Map the original movie IDs in the 'movieId' column to their
    encoded forms and store in a new column 'movie'
    df["movie"] = df["movieId"].map(movie2movie_encoded)
```

Since our training data is now ready in the form of the `users` and `movies` DataFrames, let's build the recommender model.

## Model building

In this section, we'll build the structure of our deep learning recommendation model using Keras and train it on the dataset we created in the previous section:

**Define the model**: Define the `RecommendationModel` model class, which uses embeddings for users and movies:

```
class RecommendationModel(keras.Model):
    def __init__(self, num_users, num_movies, embedding_size,
**kwargs):
        super().__init__(**kwargs)
        self.num_users = num_users
        self.num_movies = num_movies
        self.embedding_size = embedding_size
        # User embeddings layer: Represents each user as a vector in
the embedding space
        self.user_embedding = layers.Embedding(
            num_users, embedding_size,
            embeddings_initializer="he_normal",
            embeddings_regularizer=keras.regularizers.l2(1e-6),)
        self.user_bias = layers.Embedding(num_users, 1)
        # Movie embeddings layer: Represents each movie as a vector in
the embedding space
        self.movie_embedding = layers.Embedding(
            num_movies, embedding_size,
            embeddings_initializer="he_normal",
            embeddings_regularizer=keras.regularizers.l2(1e-6),)
        self.movie_bias = layers.Embedding(num_movies, 1)
```

The model calculates a match score through a dot product of user and movie embeddings.

```
# Forward pass: Given user and movie IDs, predict the rating
    def call(self, inputs):
        user_vector = self.user_embedding(inputs[:, 0])
        user_bias = self.user_bias(inputs[:, 0])
        movie_vector = self.movie_embedding(inputs[:, 1])
        movie_bias = self.movie_bias(inputs[:, 1])
        dot_user_movie = tf.tensordot(user_vector, movie_vector, 2)
        x = dot_user_movie + user_bias + movie_bias
        # The sigmoid activation forces the rating to between 0 and 1
        return tf.nn.sigmoid(x))
```

**Model compilation**: Compile the model using binary cross-entropy as the loss function and Adam as the optimizer:

```
# # Instantiate the Recommender model with the defined number of
users, movies, and embedding size
model = RecommendationModel(num_users, num_movies, EMBEDDING_SIZE)
# Compile the Recommender model
model.compile(
    #Define loss function
    loss=tf.keras.losses.BinaryCrossentropy(),
    #Define Optimizer function
 optimizer=keras.optimizers.Adam(learning_rate=0.001))
```

**Model training**: Train the model using the training data and validate it using the validation data:

```
# Train the model
history = model.fit(
    x=x_train,y=y_train,batch_size=64,
    epochs=5,verbose=1,validation_data=(x_val, y_val),)
```

## Local model testing

Before deploying the model, let's test its predictions locally to ensure it works as expected:

1.  **Prepare the test data**: For a random user, create an array of movies they haven't watched yet:

    ```
    #Load the metadata for the movies
    movie_df = pd.read_csv(movielens_dir / "movies.csv")
    # Pick a user and select their top recommendations.
    user_id = df.userId.sample(1).iloc[0]
    movies_watched_by_user = df[df.userId == user_id]
    movies_not_watched = movie_df[~movie_df["movieId"] .isin(
        movies_watched_by_user.movieId.values)]["movieId"]
    movies_not_watched = list(set(movies_not_watched).intersection(
        set(movie2movie_encoded.keys())))
    movies_not_watched = [[movie2movie_encoded.get(x)] for x in
    movies_not_watched]
    user_encoder = user2user_encoded.get(user_id)
    #Create a array of data instances to be sent for predictions
    user_prediction_array = np.hstack(([[user_encoder]] *
    len(movies_not_watched), movies_not_watched))
    ```

2. **Predict ratings**: Use the model to predict ratings for movies the user hasn't rated before:

```
#Get predicted ratings for the unwatched movies and the selected
user
ratings = model.predict(user_movie_array).flatten()
```

3. **Recommend movies**: Now, based on the predicted ratings, let's identify and display the top 10 movie recommendations for the user:

```
# Sort and pick top 10 ratings
movie_indices_top10 = ratings.argsort()[-10:][::-1]

movie_recommendations_ids = [
    movie_encoded2movie.get(movies_not_watched[x][0]) for x in
movie_indices_top10
]

print("----" * 10)
print("Top movies recommendations for user id: {}".format(user_
id))
print("----" * 10)
recommended_movies = movie_df[movie_df["movieId"].isin(
    movie_recommendations_ids)]
for row in recommended_movies.itertuples():
    print(row.title, ":", row.genres)
```

The final predictions from the model are shown here. Your list of movies will vary based on the user selected from the dataset:

```
--------------------------------
Top 10 movie recommendations
--------------------------------
Volunteers (1985) : Comedy
Emperor's New Clothes, The (2001) : Comedy
Caveman (1981) : Comedy
Juwanna Mann (2002) : Comedy
Top Secret! (1984) : Comedy
Unfaithfully Yours (1948) : Comedy
Oh, God! You Devil (1984) : Comedy
Fish Story (Fisshu sutôrî) (2009) : Comedy
Kevin Smith: Too Fat For 40 (2010) : Comedy
War Dogs (2016) : Comedy
```

So, it seems like the model is working well in the local environment and can generate movie recommendations. Now, let's deploy the model to the cloud.

# Deploying the model on Google Cloud

The first step is to deploy the model on GCP so that we can upload and register our local model on Vertex AI. Follow these steps:

1. **Register the model on Vertex AI**: To register/upload the model on Vertex AI, we need to save the core machine learning model artifacts to a **Google Cloud Storage** (**GCS**) bucket:

```
# Save the model in GCS bucket so that we can import it into
Vertex AI Model Registry
MODEL_DIR = BUCKET_URI + "/model/"
model.save(MODEL_DIR)
```

Now, we must upload the saved model to the Vertex AI Model Registry. To do this, we will need to pass the following parameters:

- `display_name`: The model's display name that will be displayed in the Vertex AI Model Registry.

- `artifact_uri`: The location of the saved model in GCS.

- `serving_container_image_uri`: The Docker image to be used as a serving container. You can use one of the images provided as part of Vertex AI or upload a custom container image to the GCP Artifact Registry. This chapter's Jupyter Notebook provides more details.

- `is_default_version`: This specifies whether this will be the default version for the model resource.

- `version_ailiases`: Alternative alias names for the model version.

- `version_description`: User description of the model version.

```
#Define service container configuration
DEPLOY_GPU, DEPLOY_NGPU = (None, None)
TF = "2.12".replace(".", "-")


if DEPLOY_GPU:
    DEPLOY_VERSION = "tf2-gpu.{}".format(TF)
else:
    DEPLOY_VERSION = "tf2-cpu.{}".format(TF)



DEPLOY_IMAGE = "{}-docker.pkg.dev/vertex-ai/prediction/
{}:latest".format(
    REGION.split("-")[0], DEPLOY_VERSION)


#Upload the Model to Vertex AI Model Registry
model = aip.Model.upload(
```

```
            display_name="recommender_model_chp15",
            artifact_uri=MODEL_DIR,
            serving_container_image_uri=DEPLOY_IMAGE,
            is_default_version=True,
            version_aliases=["v1"],
            version_description="This is the first version of the
model",)
```

2. **Deploy the model as a Vertex AI endpoint**:

   Create an endpoint on Vertex AI and deploy the model for real-time inference. Ensure you provide a display name:

   ```
   endpoint = aip.Endpoint.create(
       display_name="recommender_model_chp15",
       project=PROJECT_ID,
       location=REGION,)
   print(endpoint)
   ```

   Now, deploy the model to the newly created endpoint while specifying the machine type and other configuration settings:

   ```
   #Deploy the model to the Vertex AI endpoint
   DEPLOY_COMPUTE = "n1-standard-4" #Virtual Machine type
   response = endpoint.deploy(
       model=model,
       deployed_model_display_name="example_",
       machine_type=DEPLOY_COMPUTE,)
   print(endpoint)
   ```

## Using the model for inference

This is the fun part! We'll use our machine learning model, which has been deployed in Google Cloud Vertex AI, to make predictions using API calls. First, we'll create a Python function to send the prediction requests, then create a test/inference dataset we can send to the model as part of our request. Finally, we'll parse the prediction response we receive back from the model. Let's get started:

1. **Create a prediction function**: Create a function called `predict_custom_trained_model_sample` so that you can make predictions using the deployed model on Vertex AI:

   ```
   def predict_custom_trained_model_sample(
       project: str,
       endpoint_id: str,
       instances: Union[Dict, List[Dict]],
   ```

```
    location: str = "us-central1",
    api_endpoint: str = "us-central1-aiplatform.googleapis.
com",):
    # Initialize client that will be used to create and send
requests.
    client_options = {"api_endpoint": api_endpoint}
    client = aiplatform.gapic.PredictionServiceClient(
        client_options=client_options)
    # The format of each instance should conform to the deployed
model's prediction
    instances = [
        json_format.ParseDict(instance_dict, Value()) for
instance_dict in instances]
    parameters_dict = {}
    parameters = json_format.ParseDict(parameters_dict, Value())
    endpoint = client.endpoint_path(
        project=project, location=location, endpoint=endpoint_
id)
    response = client.predict(
        endpoint=endpoint, instances=instances,
parameters=parameters)
    print(" deployed_model_id:", response.deployed_model_id)
    # The predictions are a google.protobuf.Value representation
of the model's predictions.
    predictions = response.predictions
    return(predictions)
```

2.  **Create the inference dataset**: Create a sample inference dataset for a user and genre:

```
# Pick a random user for whom we can try to predict movie
predictions
user_id = df.userId.sample(1).iloc[0]

#Add filter for the category for which you need recommendations
genre_filter = "Drama"
```

• Create a prediction input dataset consisting of all movies in the selected genre that the user has not watched (not rated):

```
# Create Test Dataset for a User and the selected Genre
movie_df = pd.read_csv(movielens_dir / "movies.csv")

movies_watched_by_user = df[df.userId == user_id]

#Create Dataframe with Movies not watched by the User
```

```
movies_not_watched_df = movie_df[
    (~movie_df["movieId"].isin(movies_watched_by_user.movieId.
values)) & (movie_df["genres"].str.contains(genre_filter))
][["movieId","title","genres"]]


#Get the list of Movie Ids which can the be encoded using the
movie id encoder we had built earlier
movies_not_watched = movies_not_watched_df["movieId"]

movies_not_watched = list(
    set(movies_not_watched).intersection(set(movie2movie_
encoded.keys())))

movies_not_watched = [[movie2movie_encoded.get(x)] for x in
movies_not_watched]

#Get the encoded value of the user id based on the encoder built
earlier
user_encoder = user2user_encoded.get(user_id)
user_movie_array = np.hstack(([[user_encoder]] * len(movies_not_
watched), movies_not_watched))

#Create data instances that would be sent to the API for
inference
instances = user_movie_array.tolist()
```

3. **Send a prediction request**: Submit the inference dataset to the Vertex AI Prediction API endpoint:

```
# Get predicted ratings for the unwatched movies and the
selected user
predictions = predict_custom_trained_model_sample(
    project=endpoint.project,
    endpoint_id=endpoint.name,
    location=endpoint.location,
    instances = instances)
```

4. **Parse the results**: Parse the predictions received from the API endpoint and combine them:

```
# Create a DataFrame from the predictions list/array
predictions_df = pd.DataFrame(predictions)
# Rename the column in the predictions DataFrame to 'rating'
predictions_df.columns = ['rating']
```

5. **Create a DataFrame**: Create a DataFrame from the instances list/array:

```
instances_df = pd.DataFrame(instances)

# Rename the columns in the instances DataFrame to 'userId' and
```

```
'movieId' respectively
instances_df.columns = ['userId','movieId']
# Merge the instances and predictions DataFrames
combined_results = instances_df.join(predictions_df)
# Sort the results by the rating column in descending order
combined_results_sorted = combined_results.sort_
values('rating',ascending=False)
# Filter the results to show only the top 15 results
combined_results_sorted_top = combined_results_sorted.head(15)
["movieId"].values
# Map the encoded Movie IDs to the actual Movie IDs
recommended_movie_ids = [
    movie_encoded2movie.get(x) for x in combined_results_sorted_
top]
```

6. **Print**: Print the final recommended list of movies:

```
print("----" * 10)
print("Top 15 recommended movies recommendations for
User:",user_id," and Genre",genre_filter)
print("Genre:",genre_filter)
print("----" * 10)
recommended_movies = movie_df[movie_df["movieId"].
isin(recommended_movie_ids)]
for row in recommended_movies.itertuples():
    print(row.title, ":", row.genres)
```

7. The output will look something like this:

```
---------------------------------------------
Top 15 recommended movies recommendations for User: 551
Genre: Drama
---------------------------------
Stunt Man, The (1980) :
Action|Adventure|Comedy|Drama|Romance|Thriller
Affair of the Necklace, The (2001) : Drama
Baran (2001) : Adventure|Drama|Romance
Business of Strangers, The (2001) : Action|Drama|Thriller
No Man's Land (2001) : Drama|War
Blue Angel, The (Blaue Engel, Der) (1930) : Drama
Moscow on the Hudson (1984) : Comedy|Drama
Iris (2001) : Drama
Kandahar (Safar e Ghandehar) (2001) : Drama
Lantana (2001) : Drama|Mystery|Thriller
Brothers (Brødre) (2004) : Drama
Flightplan (2005) : Action|Drama|Thriller
```

```
Green Street Hooligans (a.k.a. Hooligans) (2005) : Crime|Drama
History of Violence, A (2005) : Action|Crime|Drama|Thriller
Oliver Twist (2005) : Drama
```

Starting with a dataset of movies rated by users, we were able to train a model that can now provide movie recommendations for the users in the group.

## Summary

In this chapter, we provided a brief overview of recommender systems, different techniques used for building them, and detailed steps for training, deploying, and querying a movie recommender model on Google Cloud's Vertex AI. Since the key objective was to showcase how you can address a real-world use case using GCP Vertex AI, we kept the core model somewhat simple. But if you are interested in doing a deeper dive into recommender solutions, you can look at courses such as *Recommender Systems Specialization* on Coursera.

In the next chapter, we will look into another real-world use case around building a vision-based machine learning solution to detect defects during the manufacturing process.

## References

To learn more about the topics that were covered in this chapter, take a look at the following resources:

*Collaborative Filtering for Movie Recommendations*: https://keras.io/examples/structured_data/collaborative_filtering_movielens/

*Get started with Vertex AI Model Registry*: https://github.com/GoogleCloudPlatform/vertex-ai-samples/blob/main/notebooks/official/model_registry/get_started_with_model_registry.ipynb

# 16

# Vision-Based Defect Detection System – Machines Can See Now!

**Computer Vision (CV)** is a field of artificial intelligence concerned with giving machines the ability to analyze and extract meaningful information from digital images, videos, and other visual input, as well as take actions or make recommendations based on the extracted information. Decades of research in the field of CV have led to the development of powerful **Machine Learning (ML)**-based vision algorithms that are capable of classifying images into some pre-defined categories, detecting objects from images, understanding written content from digital images, and detecting actions being performed in videos. Such vision algorithms have given businesses and organizations the ability to analyze large amounts of digital content (images and videos) and also automate processes to make instant decisions.

CV-based algorithms have changed the way we interact with smart devices in our day-to-day life – for example, we can now unlock smartphones by just showing our face, and photo editing apps today can make us look younger or older. Another important use case of applying CV-based ML algorithms is defect detection. ML algorithms can be leveraged to analyze visual input and detect defects in product images, which can be quite useful for manufacturing industries.

In this chapter, we will develop a real-world defect detection solution using deep learning on Google Cloud. We will also see how to deploy our vision-based defect detection model as a Vertex AI endpoint so that it can be utilized for online prediction.

This chapter covers the following main topics:

- Vision-based defect detection
- Deploying a vision model to a Vertex AI endpoint
- Getting online predictions from a vision model

# Technical requirements

The code samples used in this chapter can be found at the following GitHub address: `https://github.com/PacktPublishing/The-Definitive-Guide-to-Google-Vertex-AI/tree/main/Chapter16`.

# Vision-based defect detection

CV is capable nowadays of detecting visual defects on object surfaces or inconsistencies in their designs (such as dents or scratches on a car body), by just analyzing their digital photographs or videos. The manufacturing industry can leverage CV algorithms to automatically detect and remove low-quality and defected products from being packed and reaching customers. There are many possible ways to detect defects within digital content using CV-based algorithms. One simple idea is to solve defect detection as a classification problem, where a vision model can be trained to classify images such as *good* or *defected*. A more complex defect detection system will also locate the exact area of an image with a defect. The problem of identifying and locating visual defects can be solved using object-detection algorithms.

In this section, we will build and train a simple defect detection system step by step. In this example, we will use ML classification as a mechanism to detect visually defected products. Let's explore the example.

## Dataset

For this experiment, we have downloaded an open source dataset from Kaggle. The dataset has over a thousand colored images of glass bangles. These images contain bangles of different sizes and colors and can be classified into three major categories, based on their manufacturing quality and damages – good, defected, and broken. Defected bangles may have a manufacturing defect such as invariable width or improper circular shape, while broken bangles would have some piece of the circle missing. In this experiment, we will utilize some of these images to train an ML classification model and test it on a few of the unseen samples. Images are already separated into the aforementioned categories, so there's no need for manual data annotation. The dataset used in this experiment can be downloaded from the Kaggle link provided in the Jupyter Notebook corresponding to this experiment. The GitHub location of the code samples is presented at the beginning of this chapter in the *Technical requirements* section.

We have already downloaded and extracted the dataset into the same directory as our Jupyter Notebook. Now, we can start looking at some of the image samples. Let's get into the coding part now.

## Importing useful libraries

The first step is to import useful Python packages for our experiment. Here, `cv2` refers to the OpenCV library, which has lots of prebuilt functionalities for dealing with images and other CV tasks:

```
import numpy as np
import glob
import matplotlib.pyplot as plt
import cv2
from tqdm import tqdm_notebook
import tensorflow

from sklearn.model_selection import train_test_split
from sklearn.metrics import classification_report, confusion_matrix
%matplotlib inline
```

Next, let's look at the dataset.

## Loading and verifying data

Now, let's load all the image file paths into three separate lists, one for each category – good, defected, and broken. Keeping three separate lists will make it easier to keep track of image labels. Let's also print the exact number of images within each category:

```
good_bangle_paths = glob.glob("dataset/good/*.jpg")
defected_bangle_paths = glob.glob("dataset/defect/*.jpg")
broken_bangle_paths = glob.glob("dataset/broken/*.jpg")
print(len(good_bangle_paths), len(defected_bangle_paths), \
    len(broken_bangle_paths))
```

Here is the output:

```
520 244 316
```

We have 520 good-quality images, 244 defected images, and 316 broken bangle images in total. Next, let's verify a few samples from each category by plotting them using `matplotlib`.

## Checking few samples

In this step, we will randomly choose a few image paths from each of the previously discussed lists and plot them with their category name as their title.

Let's plot a few good bangle images:

```
plt.figure(figsize=(10, 10))
for ix, img_path in enumerate( \
    np.random.choice(good_bangle_paths, size=5)
):
    img = cv2.imread(img_path)
    img = cv2.cvtColor(img, cv2.COLOR_BGR2RGB)
    plt.subplot(550 + 1 + ix)
    plt.imshow(img)
    plt.axis('off')
    plt.title('Good Bangle!')
plt.show()
print("-"*101)
```

Similarly, we will plot a few random defective bangle pieces:

```
plt.figure(figsize=(10, 10))
for ix, img_path in enumerate( \
    np.random.choice(defected_bangle_paths, size=5)
):
    img = cv2.imread(img_path)
    img = cv2.cvtColor(img, cv2.COLOR_BGR2RGB)
    plt.subplot(550 + 1 + ix)
    plt.imshow(img)
    plt.axis('off')
    plt.title('Defected Bangle!')
plt.show()
print("-"*101)
```

Finally, we will also plot some broken bangle images in a similar way so that we can see all three categories visually and learn more about the data:

```
plt.figure(figsize=(10, 10))
for ix, img_path in enumerate( \
    np.random.choice(broken_bangle_paths, size=5)
):
    img = cv2.imread(img_path)
    img = cv2.cvtColor(img, cv2.COLOR_BGR2RGB)
    plt.subplot(550 + 1 + ix)
    plt.imshow(img)
    plt.axis('off')
    plt.title('Broken Bangle!')
plt.show()
```

The output of the preceding scripts is shown in *Figure 16.1* where each row represents a few random samples from each of the aforementioned categories. The category name is present in the title of images.

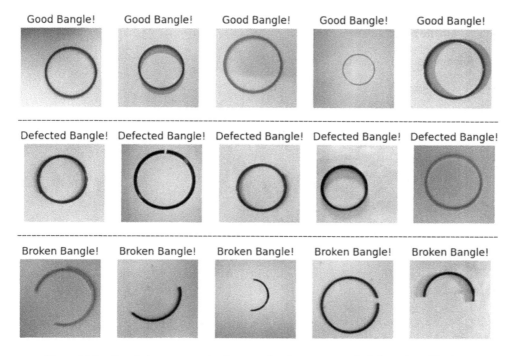

Figure 16.1 – Bangle image samples from each category – good, defected, and broken

As shown in the preceding figure, good bangles are the ones that look like a perfect circle with an even width, defected bangles may have some uneven width or surface, while broken bangles are easily distinguishable, as they are missing some part of their circular shape. So, differentiating between good and defected bangles can be a bit challenging, but differentiating both of these categories from the broken bangles should be easier for the algorithm. Let's see how it goes in the next few sections.

## Data preparation

In this step, we will prepare data for our **TensorFlow (TF)**-based deep learning model. A TF-based model requires all the inputs to have the same shape, so the first step would be to make all the input images the same shape. We will also downgrade the quality of images a bit so that they are more memory-efficient while reading them and performing calculations over them during training. We have to keep one thing in mind – we can't degrade the data quality significantly such that it becomes hard for the model to find any visual clues to detect defected surfaces. For this experiment, we will resize each image to a 200x200 resolution. Secondly, as we are only concerned about finding defects, the color of the image is not important to us. So, we will convert all the colored images into grayscale images, as it reduces the images' channels from 3 to 1; thus, the image size becomes a third of its original size. Finally, we will convert them into NumPy arrays, as TF models require the input tensors to be NumPy arrays.

The following are the pre-processing steps that we will follow for each image in the dataset:

- Reading the colored image using the OpenCV library
- Resizing an image to a fixed size (200 x 200)
- Converting image to grayscale (black and white)
- Converting the list of images into NumPy arrays
- Adding one channel dimensions, as required by convolutional layers

Let's first follow these steps for good bangle images:

```
good_bangles = []
defected_bangles = []
broken_bangles = []

for img_path in tqdm_notebook(good_bangle_paths):
    img = cv2.imread(img_path)
    img = cv2.resize(img, (200, 200))
    img = cv2.cvtColor(img, cv2.COLOR_BGR2GRAY)
    good_bangles.append(img)
good_bangles = np.array(good_bangles)
good_bangles = np.expand_dims(good_bangles, axis=-1)
```

Similarly, we will pre-process the defected bangle images:

```
for img_path in tqdm_notebook(defected_bangle_paths):
    img = cv2.imread(img_path)
    img = cv2.resize(img, (200, 200))
    img = cv2.cvtColor(img, cv2.COLOR_BGR2GRAY)
    defected_bangles.append(img)
defected_bangles = np.array(defected_bangles)
defected_bangles = np.expand_dims(defected_bangles, axis=-1)
```

Finally, we will follow the same pre-processing steps for the broken bangle images. We will also print the final shapes of NumPy arrays for each category of images:

```
for img_path in tqdm_notebook(broken_bangle_paths):
    img = cv2.imread(img_path)
    img = cv2.resize(img, (200, 200))
    img = cv2.cvtColor(img, cv2.COLOR_BGR2GRAY)
```

```
        broken_bangles.append(img)
broken_bangles = np.array(broken_bangles)
broken_bangles = np.expand_dims(broken_bangles, axis=-1)

print(good_bangles.shape, defected_bangles.shape, \
    broken_bangles.shape)
```

Each image array now will have the shape of 200x200x1, and the first dimension will represent the total number of images. The following is the output of the previous script:

```
(520, 200, 200, 1) (244, 200, 200, 1) (316, 200, 200, 1)
```

Now, let's split the data into training and test partitions.

## Splitting data into train and test

As our data is now ready and compatible with TF-model format, we just need to perform one last important step of splitting the data into two partitions – train and test. The *train* partition will be shown to the model for learning purposes during training of the model, and the *test* partition will be kept separate and will not contribute in the model parameters' update. Once the model is trained, we will check how well it performs on the unseen test partition of data.

As our test data should also have a significant number of samples from each category, we will divide each of the image categories into train and test partitions and, finally, merge all the train and test partitions together. We will utilize the first 75% of the images from each category array for training and the rest of the 25% images for testing purposes, as shown in the following snippet:

```
good_bangles_train = good_bangles[:int( \
    len(good_bangles)*0.75),]
good_bangles_test = good_bangles[int( \
    len(good_bangles)*0.75):,]
defected_bangles_train = defected_bangles[:int( \
    len(defected_bangles)*0.75),]
defected_bangles_test = defected_bangles[int( \
    len(defected_bangles)*0.75):,]
broken_bangles_train = broken_bangles[:int( \
    len(broken_bangles)*0.75),]
broken_bangles_test = broken_bangles[int( \
    len(broken_bangles)*0.75):,]

print(good_bangles_train.shape, good_bangles_test.shape)
```

The previously defined code also prints the shape of the train and test partitions just for data verification purposes. The following is the output:

```
(390, 200, 200, 1) (130, 200, 200, 1)
```

Our training and test data is almost ready now; we just need corresponding labels for model training. In the next step, we will create label arrays for both partitions.

## Final preparation of training and testing data

Now that we have three pairs of train and test partitions (one for each category), let's combine them into a single pair of train and test partitions for model training and testing purpose. After concatenating these NumPy arrays, we will also perform reshuffling so that images from each of the categories are well-mixed. This is important, as during training we will only send small batches to the model, so for smooth training of the model, each batch should have samples from all the classes. As test data is kept separate, there is no need to shuffle it.

Additionally, we also need to create corresponding label arrays as well. As ML algorithms only support numeric data, we need to encode our output categories into some numeric values. We can represent our three categories with three numbers – 0, 1, and 2.

We will use the following label mapping rule to encode our categories:

- **Good – 0**
- **Defected – 1**
- **Broken – 2**

The following code snippet concatenates all the training partitions into a single train partition and also creates a label array, using the aforementioned mapping rule:

```
all_train_images = np.concatenate((good_bangles_train, \
    defected_bangles_train, broken_bangles_train), axis=0)
all_train_labels = np.concatenate((
    np.array([0]*len(good_bangles_train)),
    np.array([1]*len(defected_bangles_train)),
    np.array([2]*len(broken_bangles_train))),
    axis=0
)
```

Similarly, we will concatenate all the test partitions into a single test partition. We will also create a label array for our test partition, which will help us to check the accuracy metrics of our trained model. Here, we also print the final shapes of the train and test partitions, as shown in the following code:

```
all_test_images = np.concatenate((good_bangles_test, \
    defected_bangles_test, broken_bangles_test), axis=0)
```

```
all_test_labels = np.concatenate((
    np.array([0]*len(good_bangles_test)),
    np.array([1]*len(defected_bangles_test)),
    np.array([2]*len(broken_bangles_test))),
    axis=0
)
print(all_train_images.shape, all_train_labels.shape)
print(all_test_images.shape, all_test_labels.shape)
```

The output of this script shows that we have 810 training images and 270 test images in total, as shown in the following output:

```
(810, 200, 200, 1) (810,)
(270, 200, 200, 1) (270,)
```

As discussed before, it is very important to shuffle our training partition, ensuring that images from each category are well-mixed and each batch will have a good variety. The important thing to keep in mind while shuffling is that we also need to shuffle the label array accordingly to avoid any data label mismatches. For this purpose, we have defined a Python function that shuffles two given arrays in unison:

```
def unison_shuffled_copies(a, b):
    assert len(a) == len(b)
    p = np.random.permutation(len(a))
    return a[p], b[p]

all_train_images, all_train_labels = unison_shuffled_copies( \
    all_train_images, all_train_labels)
all_test_images, all_test_labels = unison_shuffled_copies( \
    all_test_images, all_test_labels)
```

Our training and testing dataset is now all set. We can now move to the model architecture.

## TF model architecture

In this section, we will define a model architecture for our TF-based deep learning classification model. As we are dealing with images in this experiment, we will utilize **Convolutional Neural Network (CNN)** layers to learn and extract important features from training images. CNNs have proved to be quite useful in the field of CV. In general, a few CNN layers are stacked on top of each other to extract low-level (minor details) and high-level (big shape-related) feature information. We will also create a CNN-based feature extraction architecture in a similar way, combining it with a few fully connected layers. The final fully connected layer should have three neurons to generate output for each category, and a *sigmoid* activation layer to have that output in the form of a probability distribution.

## A convolutional block

We will now define a reusable convolutional block that we can use repeatedly to create our final model architecture. In the convolutional block, we will have each convolutional layer followed by the layers of **Batch Normalization** (**BN**), **ReLU** activation, a max-pooling layer, and a dropout layer. Here, the layers of BN and dropout are for regularization purposes to ensure the smooth learning of our TF model. The following Python snippet defines our convolutional block:

```python
def convolution_block(data, filters, kernel, strides):
    data = tensorflow.keras.layers.Conv2D(
        filters=filters,
        kernel_size=kernel,
        strides=strides,
    )(data)
    data = tensorflow.keras.layers.BatchNormalization()(data)
    data = tensorflow.keras.layers.Activation('relu')(data)
    data = tensorflow.keras.layers.MaxPooling2D(strides=strides)(data)
    data = tensorflow.keras.layers.Dropout(0.2)(data)
    return data
```

Now, let's define our complete model architecture by making use of this convolutional block.

## TF model definition

We can now define our final TF model architecture, which can utilize a few iterations of the convolutional block defined in the previous step. We will first define an input layer to tell the model about the size of the input images that it will be expecting during training. As discussed earlier, each image in our dataset has a size of 200x200x1. The following Python code defines the convolution-based feature extraction part of our network:

```python
input_data = tensorflow.keras.layers.Input(shape=(200, 200, 1))

data = input_data
data = convolution_block(data, filters=64, kernel=2, strides=2)
data = convolution_block(data, filters=128, kernel=2, strides=2)
data = convolution_block(data, filters=256, kernel=2, strides=2)
data = convolution_block(data, filters=256, kernel=2, strides=1)
```

Next, we will use a `Flatten()` layer to bring all the features into a single dimension and apply fully connected layers to refine these features. Finally, we use another fully connected layer with three neurons, followed by a `softmax` activation, to generate probabilistic output for three classes. We then define our model object with input and output layers and print out a summary of the model for reference:

```python
data = tensorflow.keras.layers.Flatten()(data)
data = tensorflow.keras.layers.Dense(64)(data)
```

```
data = tensorflow.keras.layers.Activation('relu')(data)
data = tensorflow.keras.layers.Dense(3)(data)
output_data = tensorflow.keras.layers.Activation('softmax')(data)

model = tensorflow.keras.models.Model(inputs=input_data, \
    outputs=output_data)
model.summary()
```

This snippet prints the model summary, which looks something similar to what is shown in *Figure 16.2* (for a complete summary, check out the Jupyter Notebook):

```
Model: "model"
_____
 Layer (type)                Output Shape              Param #
=================================================================
 input_1 (InputLayer)        [(None, 200, 200, 1)]     0

 conv2d (Conv2D)             (None, 100, 100, 64)      320

 batch_normalization (BatchN (None, 100, 100, 64)      256
 ormalization)

 activation (Activation)     (None, 100, 100, 64)      0

 max_pooling2d (MaxPooling2D (None, 50, 50, 64)        0
 )

 dropout (Dropout)           (None, 50, 50, 64)        0

           ●     ●     ●     ●

           ●     ●     ●     ●

 flatten (Flatten)           (None, 256)               0

 dense (Dense)               (None, 64)                16448

 activation_4 (Activation)   (None, 64)                0

 dense_1 (Dense)             (None, 3)                 195

 activation_5 (Activation)   (None, 3)                 0

=================================================================
Total params: 446,403
Trainable params: 444,995
Non-trainable params: 1,408
_____
```

Figure 16.2: The model Summary for the TF-based defect detection architecture

Our TF model graph is now ready, so we can now compile and fit our model.

## Compiling the model

In this step, we can define the appropriate loss function, optimization algorithm, and metrics. As we have a multi-label classification problem, we will utilize *categorical cross-entropy* as a loss function. We will utilize the *Adam* optimizer with its default values and `'accuracy'` as a metric:

```
model.compile(
    loss='sparse_categorical_crossentropy',
    optimizer='adam',
    metrics=['accuracy']
)
```

Now, we are all set to start training our model on the previously curated dataset.

## Training the model

Now, we are all set to launch the training of our model, as our data and model object are both set. We plan to train our model for 50 epochs with a batch size of 64. After each epoch, we will keep checking the model's loss and accuracy on training and testing partitions:

```
history = model.fit(
    x=all_train_images,
    y=all_train_labels,
    batch_size=64,
    epochs=50,
    validation_data=(all_test_images, all_test_labels),
)
```

The training logs with loss and accuracy values will look similar to the following:

```
Epoch 1/50
13/13 [==============================] - 6s 311ms/step - loss: 1.1004
- accuracy: 0.5000 - val_loss: 7.9670 - val_accuracy: 0.2259
Epoch 2/50
13/13 [==============================] - 3s 260ms/step - loss: 0.8947
- accuracy: 0.5938 - val_loss: 11.9169 - val_accuracy: 0.2259
-     -    -    -
Epoch 50/50
13/13 [==============================] - 3s 225ms/step - loss: 0.1369
- accuracy: 0.9481 - val_loss: 0.9686 - val_accuracy: 0.6741
```

Once the training is complete, we can start verifying the results of our model.

## Plotting the training progress

In this step, we will utilize the `history` variable defined in the previous step to plot the progress of the training loss, test loss, train accuracy, and test accuracy with progressing epochs. These graphs can help us understand whether our model training is going in the right direction or not. Also, we can check the ideal number of epochs required to get reasonable accuracy on test sets.

Let's first plot the training and validation loss of the model with an epoch number on the $X$ axis and a loss value on the $Y$ axis:

```
plt.plot(history.history['loss'])
plt.plot(history.history['val_loss'])
plt.title('model loss')
plt.ylabel('loss')
plt.xlabel('epoch')
plt.legend(['train', 'val'], loc='upper left')
plt.show()
```

The output of this snippet is shown in *Figure 16.3*. We can see in the figure that training and validation loss decrease as the training progresses, which tells us that our model training is going in the right direction and our model is learning.

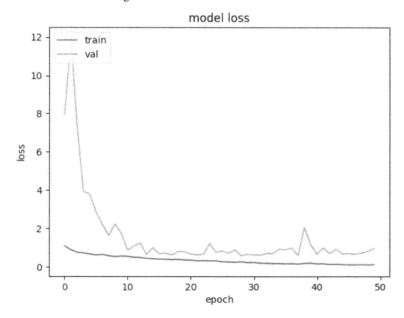

Figure 16.3 – Training and validation loss

Next, we can also plot the accuracy of the model on training and test partitions with the progress of training, as shown in following code:

```
plt.plot(history.history['accuracy'])
plt.plot(history.history['val_accuracy'])
plt.title('TF model accuracy trend.')
plt.ylabel('TF model accuracy')
plt.xlabel('epoch number')
plt.legend(['train', 'val'], loc='upper left')
plt.show()
```

The resulting plot can be seen in *Figure 16.4*. We can see that training accuracy keeps increasing and reaches close to 100% as the model starts overfitting, while the validation accuracy increases to around 70% and keeps fluctuating around that. It means that our current setup is capable of achieving around 70% accuracy on the test set. This accuracy value can be improved further by either increasing the capacity of our network (by adding a few more layers), or by improving the way we extract features from the images. For our experiment, this accuracy is satisfactory, and we will move forward with it.

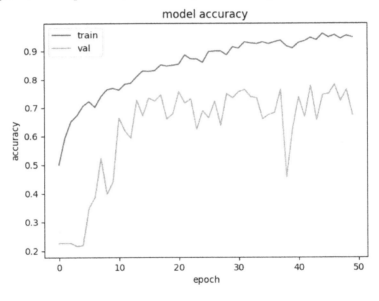

Figure 16.4 – Training and validation accuracy with the progress of training epochs

As our model training is now complete, we can start making predictions on unseen data. Let's first check the results on our test set.

## Results

Here comes the interesting part where we check the TF-model results on our unseen test set. Before checking the numbers, let's first plot a few test images along with their true labels and model outputs.

## Checking the results on a few random test images

Let's first visually verify the results of the model by choosing a few random images from our test set, making predictions on them, and plotting them with model predictions along with actual labels. We will put the label and prediction information in the image titles, as shown in the following Python snippet:

```
for iteration in range(2):
    plt.figure(figsize=(12, 12))
    for idx, img_idx in enumerate(np.random.permutation( \
        len(all_test_images))[:4]
    ):
        img = all_test_images[img_idx]
        label = all_test_labels[img_idx]
        pred = model.predict(np.array([img]), verbose=0)
        model_output = np.argmax(pred)
        plt.subplot(440 + 1 + idx)
        plt.imshow(img, cmap='gray')
        plt.title(f'Label: {label}, Model_output: \
            {model_output}')
        plt.axis('off')
    plt.show()
```

The output of this snippet is shown in *Figure 16.4*. We can see that the model is slightly confused between class 0 and class 1 but easily identifies class 2, due to big clues on the shape. As a reminder, class 0 here represents good bangles, class 1 represents defected bangles, and class 2 represents broken bangles.

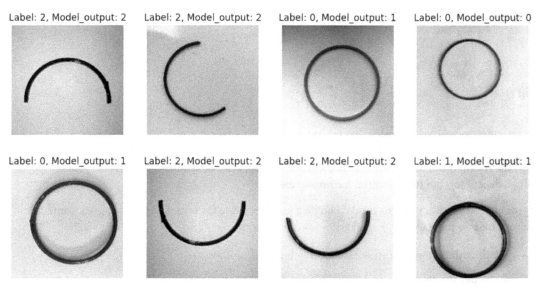

Figure 16.5 – The model results for the classification of good, defected, and broken bangles

As per *Figure 16.5*, our model is doing a good job of identifying class 2 but is slightly confused between class 0 and class 1, due to very tiny visual clues. Next, let's check the metrics on the entire test set to get a sense of the quality of our classifier.

## Classification report

To generate the classification report on our entire test set, we first need to generate model outputs for the entire test set. We will choose the class with maximum probability as the model prediction (note that choosing the output class like that may not be the best option when we have highly imbalanced datasets). The following is the Python code that generates the model output on the entire test set:

```
test_pred = model.predict(all_test_images)
test_outputs = [np.argmax(pred) for pred in test_pred]
```

We can now print the classification report for our model:

```
print(
    classification_report(all_test_labels,
                   test_outputs,
                   target_names=['Good', 'Defected', 'Broken'],
    )
)
```

The following is the output of the classification report:

|  | precision | recall | f1-score | support |
|---|---|---|---|---|
| Good | 0.88 | 0.52 | 0.65 | 130 |
| Defected | 0.62 | 0.69 | 0.65 | 61 |
| Broken | 0.58 | 0.92 | 0.71 | 79 |
|  |  |  |  |  |
| accuracy |  |  | 0.67 | 270 |
| macro avg | 0.69 | 0.71 | 0.67 | 270 |
| weighted avg | 0.73 | 0.67 | 0.67 | 270 |

The classification report indicates that our model has a F1 score of around 0.65 for class 0 and class 1, and 0.71 for class 2. As suspected, the model is doing a better job at identifying broken bangle images. Thus, the recall of the model for the Broken class is very good, around 92%. Overall, our model is doing a decent job but has potential for improvement.

Improving the accuracy of this model can be a good exercise for you. The following are some hints that may help to increase the overall accuracy of the model:

* Work with better resolution (something better than 200x200)

* A deeper model (more and bigger CNN layers to get better features)

- Data augmentation (make the model more robust)
- A better network (better feature extraction with an attention mechanism or any other feature extraction strategy)

Finally, let's print the confusion matrix:

```
confusion_matrix(all_test_labels, test_outputs,)
```

Here is the output:

```
array([[67, 21, 42],
       [ 8, 42, 11],
       [ 1,  5, 73]])
```

A confusion matrix can help to determine what kind of mistakes our model makes. In other words, when it classifies class 0 incorrectly, which class does our model confuse it with?

With this, our exercise of training a custom TF-based model for the task of defect detection is complete. We now have an average-performing trained TF model. Now, let's see how can we deploy this model on Google Vertex AI.

# Deploying a vision model to a Vertex AI endpoint

In the previous section, we completed our experiment of training a TF-based vision model to identify detects from product images. We now have a trained model that can identify defected or broken bangle images. To make this model usable in downstream applications, we need to deploy it to an endpoint so that we can query that endpoint, getting outputs for new input images on demand. There are certain things that are important to consider while deploying a model, such as expected traffic, expected latency, and expected cost. Based on these factors, we can choose the best infrastructure to deploy our models. If there are strict low-latency requirements, we can deploy our model to machines with accelerators (such as **Graphical Processing Units** (**GPUs**) or **Tensor Processing Units** (**TPUs**)). Conversely, we don't have the necessity of online or on-demand predictions, so we don't need to deploy our model to an endpoint. Offline batch-prediction requests can be handled without even deploying the model to an endpoint.

## Saving model to Google Cloud Storage (GCS)

In our case, we are interested in deploying our model to a Vertex AI endpoint, just to see how it works. The very first step is to save our trained TF model into GCS:

```
model.save(
    filepath='gs://my-training-artifacts/tf_model/',
    overwrite=True,
)
```

Once the model is saved, the next step would be to upload this model artifact to the Vertex AI Model Registry. Once our model is uploaded to the Model Registry, it can be deployed easily, either by using the Google Cloud console UI or Python scripts.

## Uploading the TF model to the Vertex Model Registry

In this step, we will utilize the Vertex AI SDK to upload our TF model to the Model Registry. Alternatively, we can also use the console UI to do the same. For this purpose, we will be required to provide a model display name, a region to upload the model to, the URI of the saved model artifact, and a serving container image. Note that the serving container images must have dependencies installed for appropriate versions of frameworks, such as TF. Google Cloud provides a list of prebuilt serving containers that can be readily used to serve the models. In our case, we will also use a prebuilt container that supports TF 2.11 for serving.

Let's set up some configurations:

```
PROJECT_ID='417812395597'
REGION='us-central1'
ARTIFACT_URI='gs://my-training-artifacts/tf_model/'
MODEL_DISPLAY_NAME='tf-bangle-defect-detector-v1'
SERVING_IMAGE='us-docker.pkg.dev/vertex-ai/prediction/tf2-cpu.2-
11:latest'
```

Now, we can go ahead and upload the model artifact to the Model Registry, as shown in the following Python code:

```
aiplatform.init(project=PROJECT_ID, location=REGION)
model = aiplatform.Model.upload(
    display_name=MODEL_DISPLAY_NAME,
    artifact_uri=ARTIFACT_URI,
    serving_container_image_uri=SERVING_IMAGE,
    sync=True,
)
model.wait()
print("Model Display Name: ", model.display_name)
print("Model Resource Name: ", model.resource_name)
```

Here is the output:

```
Model Display Name:  tf-bangle-defect-detector-v1
Model Resource Name:  projects/417812395597/locations/us-
central1/models/3991356951198957568
```

We have now successfully uploaded our TF model to the Registry. We can now locate our model, using either the model display name or the model resource ID, as per the previous output. Next, we need an endpoint to deploy our model.

## Creating a Vertex AI endpoint

In this step, we will create a Vertex AI endpoint that will be used to serve our model prediction requests. Again, the endpoints can also be created using the Google Cloud console UI, but here, we will do it programmatically with the Vertex AI SDK.

Here is the sample function that can be leveraged to create an endpoint:

```
def create_vertex_endpoint(
    project_id: str,
    display_name: str,
    location: str,
):
    aiplatform.init(project=project_id, location=location)
    endpoint = aiplatform.Endpoint.create(
        display_name=display_name,
        project=project_id,
        location=location,
    )
    print("Endpoint Display Name: ", endpoint.display_name)
    print("Endpoint Resource Name: ", endpoint.resource_name)
    return endpoint
```

Let's use this function to create an endpoint for our model:

```
ENDPOINT_DISPLAY_NAME='tf-bangle-defect-detector-endpoint'
vertex_endpoint = create_vertex_endpoint(
    project_id=PROJECT_ID,
    display_name=ENDPOINT_DISPLAY_NAME,
    location=REGION,
)
```

This snippet gives us the following output, and it confirms that the endpoint has been created (note that this endpoint is currently empty without any model):

```
Endpoint Display Name:  tf-bangle-defect-detector-endpoint
Endpoint Resource Name:  projects/417812395597/locations/us-central1/
endpoints/4516901519043330048
```

We can use the following command to list all the endpoints within a region to verify whether it has been successfully created:

```
gcloud ai endpoints list --region={REGION}
```

This command gives the following output:

```
ENDPOINT_ID            DISPLAY_NAME
4516901519043330048    tf-bangle-defect-detector-endpoint
```

Our endpoint is now set, so we can go ahead and deploy a model now.

## Deploying a model to the Vertex AI endpoint

We now have our TF model in the Model Registry, and we have also created a Vertex AI endpoint. We are now all set to deploy our model to this endpoint. This action can also be performed using the Google Cloud console UI, but here, we will do it programmatically using the Vertex AI SDK. First, let's get the details of our model from the Registry:

```
## List Model versions
models = aiplatform.Model.list( \
    filter=f"display_name={MODEL_DISPLAY_NAME}")
print("Number of models:", len(models))
print("Version ID:", models[0].version_id)
Output:
Number of models: 1
Version ID: 1
```

Now, let's deploy our model to the endpoint:

```
MODEL = models[0]
DEPLOYED_MODEL_DISPLAY_NAME='tf-bangle-defect-detector-deployed-v1'
MACHINE_TYPE='n1-standard-16'
#Deploy the model to the Vertex AI endpoint
response = vertex_endpoint.deploy(
    model=MODEL,
    deployed_model_display_name=DEPLOYED_MODEL_DISPLAY_NAME,
    machine_type=MACHINE_TYPE,
)
```

As soon as the execution of this snippet is complete without any errors, our model should be ready to accept requests for online predictions. This deployed model information should be visible in the Google Cloud console UI as well. We can also verify the deployed model by using the following function:

```
vertex_endpoint.gca_resource.deployed_models[0]
```

This should print the resources related to the deployed model. The output will look something similar to the following:

```
id: "4885783272115666944"
model: "projects/417812395597/locations/us-central1/
models/3991356951198957568"
display_name: "tf-bangle-defect-detector-deployed-v1"
create_time {
    seconds: 1694242166
    nanos: 923666000
}
dedicated_resources {
    machine_spec {
        machine_type: "n1-standard-16"
    }
    min_replica_count: 1
    max_replica_count: 1
}
model_version_id: "1"
```

We have now verified that our TF model is successfully deployed. Now, we are all set to call this endpoint for predictions. In the next section, we will see how an endpoint with a deployed model can serve online predictions.

## Getting online predictions from a vision model

In the previous section, we deployed our custom TF-based model to a Vertex AI Endpoint so that we could embed it into any downstream application, querying it for on-demand or online predictions. In this section, we will see how we can call this endpoint for online predictions programmatically using Python. However, the prediction requests can also be made by using a `curl` command and sending a JSON file with input data.

There are a few things to consider while making prediction requests; the most important part is pre-processing the input data accordingly. In the first section, when we trained our model, we did some pre-processing on our image dataset to make it compatible with the model. Similarly, while requesting the predictions, we should follow the exact same data preparation steps. Otherwise, either the model request will fail, due to an incompatible input format, or it will give bad results, due to training-serving skew. We already have pre-processed test images, so we can pass them directly within the prediction requests.

As we will pass the input image to the request JSON, we will need to encode it into a JSON serializable format. So, we will send our input image to the model encode with the `base64` format. Check out the following snippet for an example payload creation of a single test image:

```
import base64
Instances = [
    {
      "input_1": {
        "b64": base64.b64encode(all_test_images[0]).decode("utf-8")
      }
    }
]
```

Let's set some configurations to call the endpoint:

```
from google.protobuf import json_format
from google.protobuf.struct_pb2 import Value
ENDPOINT_ID="4516901519043330048"
client_options = {"api_endpoint": "us-central1-aiplatform.googleapis.
com"}
client = aiplatform.gapic.PredictionServiceClient( \
    client_options=client_options)
instances = [
    json_format.ParseDict(instance_dict, Value()) for \
        instance_dict in Instances
]
parameters_dict = {}
parameters = json_format.ParseDict(parameters_dict, Value())
```

We can now go ahead and make a prediction request to our Vertex AI endpoint. Note that our parameters dictionary is empty, which means that we will get the raw model predictions as a result. In a more customized setting, we can also pass some parameters, such as *thresholds*, to perform minor post-processing on the model predictions accordingly. Check out the following Python code to request predictions:

```
endpoint = client.endpoint_path(
    project=PROJECT_ID, location=REGION, endpoint=ENDPOINT_ID
)
response = client.predict(
    endpoint=endpoint, instances=instances, \
    parameters=parameters
)
print("response")
print(" deployed_model_id:", response.deployed_model_id)
predictions = response.predictions
```

Similarly, we can pass a number of images in the `instances` variable and get an on-demand or online prediction result from an endpoint.

## Summary

In this chapter, we created an end-to-end vision-based solution to detect visual defects from images. We saw how CNN-based deep learning architectures can be used to extract useful features from images and then use those features for tasks such as classification. After training and testing our model, we went ahead and deployed it to a Vertex AI endpoint, allowing it to serve online or on-demand prediction requests for any number of downstream applications.

After completing this chapter, you should be confident about how to approach vision-based problems and how to utilize ML to solve them. You should now be able to train your own vision-based classification models to solve real-world business problems. After completing the second section on deploying a custom model to a Vertex AI endpoint and the third section on getting online prediction from a Vertex endpoint, you should now be able to make your custom vision models usable for any downstream business application, by deploying them to Google Vertex AI. We hope this chapter was a good learning experience, with a hands-on real-world example. The next chapter will also present a hands-on example of a real-world, NLP-related use case.

# 17

# Natural Language Models – Detecting Fake News Articles!

A significant amount of content on the internet is in textual format. Almost every organization stores lots of internal data and resources as text documents. **Natural language processing** (**NLP**) is a subfield of machine learning that's concerned with organizing, understanding, and making decisions based on textual input data. Over the past decade, NLP has become the utmost important aspect of transforming business processes and making informed decisions. For example, a sentiment analysis model can help a business understand the high-level sentiments of their customers toward their products and services. A topic modeling algorithm combined with sentiment analysis can figure out the key pain points of the customers and thus it can inform the business decisions to make customer satisfaction a priority.

In this chapter, we will develop an ML system that can recognize fake news articles. Such systems can help in keeping the information and news on the internet more accurate and safer. We will cover the following main topics:

- Detecting fake news using NLP
- Launching model training on Vertex AI
- BERT-based fake news classification

## Technical requirements

The code samples used in this chapter can be found in this book's GitHub repository: `https://github.com/PacktPublishing/The-Definitive-Guide-to-Google-Vertex-AI/tree/main/Chapter17`.

# Detecting fake news using NLP

Nowadays, due to the increase in the use of the internet, it has become really easy to spread fake news. A large number of users are consuming and posting content on the internet via their social media accounts daily. It has become difficult to distinguish the real news from the fake news. Fake news, however, can do significant damage to a person, society, organization, or political party. Looking at the scale, it is impossible to skim through every article manually or using a human reviewer. Thus, there is a need to develop smart algorithms that can automatically detect fake news articles and stop the spread of dangerous news as soon as it is generated.

ML-based classification algorithms can be used to detect fake news. First, we need a good training dataset to train the classification model on so that it can learn the common patterns of fake news and thus automatically distinguish it from real news. In this section, we will train an ML model to classify articles as "fake" versus "real."

## Fake news classification with random forest

In this section, we will use a tree-based classification algorithm known as random forest to detect fake news articles. In the last section of this chapter, we will also train a complex deep learning-based classifier and compare the accuracy of both models. Let's dive into the experiment. All the code related to these experiments can be found in this book's GitHub repository, as mentioned in the Technical requirements section.

## About the dataset

We have downloaded the dataset from Kaggle, which has an open-to-use license. The dataset contains about 72k news articles with titles, texts, and labels. Almost 50% of the articles are "fake," while the remainder are "real." We will utilize this dataset to train an NLP-based classification model that can detect fake news. We will keep some parts of this dataset as unseen data so that we can test the model results after training. The link for downloading the data can be found in the Jupyter Notebook in this book's GitHub repository.

> **Note**
> We have already downloaded and decompressed the data in the same directory as the Jupyter Notebook.

Now, let's jump into the implementation part. We will start by importing useful Python libraries.

## Importing useful libraries

The first step is to load some useful Python libraries in a notebook cell:

```
import numpy as np
import pandas as pd
import matplotlib.pyplot as plt
from tqdm import tqdm_notebook

from sklearn.model_selection import train_test_split
from sklearn.metrics import classification_report, confusion_matrix
from sklearn.feature_extraction.text import TfidfVectorizer
from sklearn.ensemble import RandomForestClassifier
from sklearn.metrics import roc_curve
%matplotlib inline
```

Next, we will load and verify the input dataset.

## Reading and verifying the data

Here, we will read the data from a CSV file into a pandas DataFrame called `news_df`. We will print the shape of the DataFrame and a few top entries:

```
news_df = pd.read_csv("WELFake_Dataset.csv")
print(news_df.shape)
news_df.head()
```

The output of this cell is shown in *Figure 17.1*. As we can see, there are 72,134 news articles in this table, each with a title, body, and label:

```
(72134, 4)
```

|   | Unnamed: 0 | title | text | label |
|---|---|---|---|---|
| 0 | 0 | LAW ENFORCEMENT ON HIGH ALERT Following Threat... | No comment is expected from Barack Obama Membe... | 1 |
| 1 | 1 | NaN | Did they post their votes for Hillary already? | 1 |
| 2 | 2 | UNBELIEVABLE! OBAMA'S ATTORNEY GENERAL SAYS MO... | Now, most of the demonstrators gathered last ... | 1 |
| 3 | 3 | Bobby Jindal, raised Hindu, uses story of Chri... | A dozen politically active pastors came here f... | 0 |
| 4 | 4 | SATAN 2: Russia unvelis an image of its terrif... | The RS-28 Sarmat missile, dubbed Satan 2, will... | 1 |

Figure 17.1 – Fake news article detection dataset overview

Now, let's see if there are any missing values present in this data table.

## NULL value check

We need to check if there are any NULL values present in the dataset. There are different ways of handling NULL values. If the percentage of NULL values is very low, we can choose to drop those rows from the table; otherwise, we can fill those entries with some value. In our case, we will fill the NULL fields using an empty string value:

```
news_df.isnull().sum()
```

Here's the output of this cell:

```
Unnamed: 0        0
title           558
text             39
label             0
dtype: int64
```

As we can see, there's a very small number of entries with NULL values. Let's fill them with empty strings:

```
news_df.fillna('', inplace=True)
```

We can now move on to data cleaning and pre-processing.

## Combining title and text into a single column

Let's create a new column called `content` with combined `title` and `text` elements so that it contains all the textual information available related to the news article. Once we've done this, we will be able to use this column for model training and classification purposes:

```
news_df['content'] = [x + ' ' + y for x,y in zip(news_df.title, news_df.text)]
```

Now that our textual content is in a single column, we can start cleaning and preparing it for the model.

## Cleaning and pre-processing data

ML algorithms are very sensitive to noisy data. So, it is of utmost importance to clean and process the data before passing it into the model for training; this will allow the model to learn useful information from it. As we are using a classical ML algorithm here, we will need to do some aggressive cleaning and pre-processing. In the case of deep learning, data processing is not required (as shown in the last section of this chapter). When we solve NLP problems using classical ML algorithms, we often use feature extraction methods such as TF and TF-IDF. As we know, these feature extraction methods are sensitive to the count of words, so it becomes important to remove less meaningful words (such as stopwords) and characters from the text.

In this experiment, we will follow these steps to clean and pre-process the data:

1.  Remove special characters and numbers from the text.

2.  Convert the text into lowercase (so that "HELLO" and "hello" are the same for the classification algorithm).

3.  Split the content by space to get a list of words.

4.  Remove stopwords. These are common English words and are often meaningless in a sentence. Examples include they, the, and, he, and him.

5.  Apply stemming. This involves reducing the words to their root forms (for example, "happiness" should be reduced to "happy" so that different variations of the same word are equal for the model).

Join words with spaces in between to create text.

We will require some NLP-specific libraries that will help in preparing the data. Here, we will utilize the `nltk` (Natural Language Toolkit) library to remove stopwords and apply the stemming operation. To convert our text data into a numerical format, we will utilize the `TfidfVectorizer` method from the `sklearn` library:

```
import re
from nltk.corpus import stopwords
from nltk.stem.porter import PorterStemmer
from sklearn.feature_extraction.text import TfidfVectorizer
```

> **Note**
>
> When we install the `nltk` library, it doesn't automatically download all the required resources related to it. In our case, we will have to explicitly download the English language stopwords. We can do that by running the following command in a terminal:

```
nltk.download("stopwords")
```

Here is our data cleaning and pre-processing function. As this function runs on the entire dataset, it takes some time to complete:

```
def clean_and_prepare_content(text):
    text = re.sub('[^a-zA-Z]',' ', text)
    text = text.lower()
    text_words = text.split()
    imp_text_words = [word for word in text_words if not word in
stopwords.words('english')]
    stemmed_words = [porter_stemmer.stem(word) for word in imp_text_
words]
    processed_text = ' '.join(stemmed_words)
```

```
        return processed_text

porter_stemmer = PorterStemmer()
news_df['processed_content'] = news_df.content.apply(lambda content:
clean_and_prepare_content(content))
```

Now, let's separate our content and labels into arrays for modeling purposes.

## Separating the data and labels

Here, we are separating the data and labels and putting them into two separate lists:

```
X = news_df.processed_content.values
y = news_df.label.values
print(X.shape, y.shape)
Here's the output:
(72134,) (72134,)
```

Now, let's convert the text into numeric values.

## Converting text into numeric data

As ML algorithms only understand numbers, we will need to convert the textual data into numeric format. In our experiment, we will be creating TF-IDF features:

```
vectorizer = TfidfVectorizer()
vectorizer.fit(X)
X = vectorizer.transform(X)
print(X.shape)
Here's the output:
(72134, 162203)
```

We can now split the data into training and test partitions so that we can test the results of our model after training.

> **Note**
>
> In a real NLP project, We must split the dataset into training and test sets before applying numerical transformations. The transformation function (such as `TfidfVectorizer`) should only be fit to the training data and then applied to test data. This is because, in a real-world setting, we might get some unknown words in the dataset and our model is not supposed to see those words during training. Another issue with this setting is that it causes data leakage as the statistics that are calculated over the entire dataset also belong to the test partition. In this example, we have done this transformation before splitting the dataset just for simplicity.

## Splitting the data

Next, we must split the data into training and test partitions. We will use about 80% of the data for training and the remaining 20% for testing:

```
X_train, X_test, y_train, y_test = train_test_split(X, y, test_size =
0.2, stratify=y, random_state=42)
```

Our training and test data partitions are now ready to be fed to the model. Next, we'll define the model.

## Defining the random forest classifier

For our simple experiment, we are using default hyperparameter values for the random forest model. However, in a real-world use case, we can experiment with different sets of hyperparameter values to get the best results. Alternatively, we can utilize hyperparameter tuning to find the best hyperparameters for our model:

```
rf_model = RandomForestClassifier()
```

Let's go ahead and train the model on the training partition.

## Training the model

Let's fit our model to the training dataset:

```
rf_model.fit(X_train, y_train)
```

Here's the output:

```
RandomForestClassifier()
```

Our model training is now complete, which means we can start predicting the test data to check the model's results.

## Predicting the test data

Here, we're using our trained random forest classifier to make predictions on the test partition. The `predict` function gives the class-level output, while the `predict_proba` function gives the probabilistic outputs:

```
y_pred = rf_model.predict(X_test)
y_proba = rf_model.predict_proba(X_test)
```

Here, we have made predictions on the entire set. Let's check how well our model did.

## Checking the results/metrics on the test dataset

The next important step is to check and verify the performance of our model on the test dataset. Here, we will use sklearn's classification report method to get the precision, recall, and F1 score for each class. Check out the following code snippet:

```
# classification report
print(
    classification_report(
        y_test,
        y_pred,
        target_names=['Real', 'Fake'],
    )
)
```

Here is the output of the classification report:

|  | precision | recall | f1-score | support |
|---|---|---|---|---|
| Real | 0.94 | 0.92 | 0.93 | 7006 |
| Fake | 0.93 | 0.94 | 0.94 | 7421 |
|  |  |  |  |  |
| accuracy |  |  | 0.93 | 14427 |
| macro avg | 0.93 | 0.93 | 0.93 | 14427 |
| weighted avg | 0.93 | 0.93 | 0.93 | 14427 |

As we can see, our model has about 93% precision and recall for both classes. The overall accuracy is also about 93%. So, we can say that our model is good enough to identify about 93% of the fake news articles.

Next, let's plot the ROC curve. The ROC curve is a graph between the **false positive rate (FPR)** and **true positive rate (TPR)** of a classification model:

```
def plot_roc_curve(y_true, y_prob):
    """
    plots the roc curve based of the probabilities
    """
    fpr, tpr, thresholds = roc_curve(y_true, y_prob)
    plt.plot(fpr, tpr)
    plt.xlabel('False Positive Rate')
    plt.ylabel('True Positive Rate')

plot_roc_curve(y_test, y_proba[:,1])
```

Check out *Figure 17.2* to see the ROC curve for our experiment. In a typical ROC curve, the X-axis represents the FPR and the Y-axis represents the TPR of the model. The **area under the ROC curve**, also known as **ROC-AUC**, indicates the quality of a classification model. Having a higher area value signifies a better model:

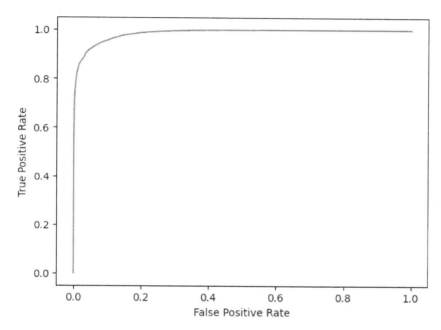

Figure 17.2 – The ROC curve for the fake news classification model

Let's also check out the confusion matrix to see where our model is making mistakes.

## Confusion matrix

Finally, let's also print the confusion matrix for our classification. A confusion matrix shows the number of correct and incorrect classifications of each class. It also shows which other classes were predicted as mistakes if the classification is wrong (false positives and false negatives):

```
confusion_matrix(y_test, y_pred,)
```

Here's the output:

```
array([[6455,  551],
       [ 409, 7012]])
```

Our experiment is now complete. If the results are satisfactory, we can go ahead and deploy this model as an API. If the results are still not acceptable, we can do more experiments with different settings.

If we want to do a lot of experiments in parallel, we can launch many parallel experiments via Vertex AI training jobs without needing to monitor them constantly and check back later when training is complete. In the next section, we will see how Vertex AI training jobs can be configured.

# Launching model training on Vertex AI

In this section, we will launch our training experiment as a Vertex AI training job. There are multiple advantages of launching training jobs on Vertex AI instead of doing it in a Juypter Notebook:

- The flexibility to launch any number of parallel experiments

- We can choose the best hardware for model training, which is very important when accelerators are needed to train deep learning models.

- We don't need active monitoring regarding training progress

- There's no fear of the Jupyter Notebook crashing

- Vertex AI training jobs can be configured to log metadata and experiments in the Google Cloud Console UI

- In this section, we will create and launch a Vertex AI training job for our experiment. There are two main things we need to do to launch a Vertex AI training job. First, we need to put the dataset in a location that will be accessible to the Vertex AI job (such as GCS or BigQuery). Second, we need to put the model training code together into a single `task.py` file so that it can be packaged into a training container with all the necessary dependencies.

Here are the steps we need to follow to create and launch our Vertex AI training job:

1. Upload the dataset to GCS or BigQuery (we will use GCS).

2. Create a `task.py` file that does the following:

   - Reads data from GCS

   - Does the necessary data preparation

   - Trains the RF model

   - Saves the trained model into GCS

   - Does prediction on the test set

   - (Optionally) Saves predictions to GCS

   - Prints some results/metrics

3. Use a prebuilt training image.

4. Launch Vertex AI training.

5. Monitor the progress on the Google Cloud Console UI.

Considering these steps, we have already created a `task.py` file for our experiment; it can be found in this book's GitHub repository. Next, we will learn how to launch the job using this `task.py` file.

## Setting configurations

Here, we will define the configurations related to the project and data locations that will be necessary while launching the training job on Vertex AI. The following snippet shows some configurations related to our experiment:

```
PROJECT_ID='417xxxxxxx97'
REGION='us-west2'
BUCKET_URI='gs://my-training-artifacts'
DATA_LOCATION='gs://my-training-artifacts/WELFake_Dataset.csv'

# prebuilt training containers
TRAIN_VERSION = "tf-cpu.2-9"
TRAIN_IMAGE = "us-docker.pkg.dev/vertex-ai/training/{}:latest".
format(TRAIN_VERSION)
```

Let's initialize the Vertex AI SDK with appropriate variables.

## Initializing the Vertex AI SDK

Here, we're initializing the Vertex AI SDK to set the project, location, and staging bucket for our jobs:

```
from google.cloud import aiplatform
aiplatform.init(project=PROJECT_ID, location=REGION, staging_
bucket=BUCKET_URI)
```

Now that our configurations have been set, we can start defining the Vertex AI training job.

## Defining the Vertex AI training job

The following code block defines a Vertex AI training job for our experiment. Here, we pass `display_name`, which will help us locate our job within the console UI. Note that we are passing our `task.py` file as the script path variable. `container_uri` is the prebuilt container that will be used to launch the job. Finally, we can specify any additional Python packages that are required to run our training code. In our case, we need to install the `nltk` package for some NLP-related functionalities:

```
job = aiplatform.CustomTrainingJob(
    display_name="fake_news_detection",
    script_path="task.py",
    container_uri=TRAIN_IMAGE,
    requirements=["nltk"],
)
```

Our Vertex AI-based custom training job is now ready. Let's run it.

## Running the Vertex AI job

We are all set to launch our training job. We are using an `n1-standard-16` type of machine for our experiment that can be modified as per our needs. Check out the following snippet, which launches our training job on Vertex AI:

```
# Start the training job
model = job.run(
    machine_type = "n1-standard-16",
    replica_count=1,
)
```

After launching the job, we should see a URL in the output pointing to the job within the Cloud Console UI. The output should look something like this:

```
Training script copied to:
gs://my-training-artifacts/aiplatform-2023-09-04-04:41:36.367-
aiplatform_custom_trainer_script-0.1.tar.gz.
Training Output directory:
gs://my-training-artifacts/aiplatform-custom-traini
ng-2023-09-04-04:41:36.625
View Training:
https://console.cloud.google.com/ai/platform/locations/us-west2/
training/8404xxxxxxxxxx898?project=417xxxxxxxx7
CustomTrainingJob projects/417xxxxxxxx7/locations/us-west2/
trainingPipelines/840xxxxxxxxxx92 current state:
PipelineState.PIPELINE_STATE_RUNNING
View backing custom job:
https://console.cloud.google.com/ai/platform/locations/us-west2/
training/678xxxxxxxxxxx48?project=417xxxxxxxx7
CustomTrainingJob projects/417xxxxxxxx7/locations/us-west2/
trainingPipelines/840xxxxxxxxxx92 current state:
PipelineState.PIPELINE_STATE_RUNNING
```

With that, we have successfully launched our experiment as a training job on Vertex. We can now monitor the progress of our job using the Cloud Console UI. Next, we'll solve this problem using a deep learning approach so that we hopefully get better results.

## BERT-based fake news classification

In our first experiment, we trained a classical random forest classifier on TF-IDF features to detect fake versus real news articles and got an accuracy score of about 93%. In this section, we will train a deep learning model for the same task and see if we get any accuracy gains over the classical tree-based approach. Deep learning has changed the way we used to solve NLP problems. Classical approaches required hand-crafted features, most of which were related to the frequency of words appearing in a document. Looking at the complexity of languages, just knowing the count of words in a paragraph is not enough. The order in which words occur also has a significant impact on the overall meaning of the paragraph or sentence. Deep learning approaches such as **Long-Short-Term-Memory (LSTM)** also consider the sequential dependency of words in sentences or paragraphs to get a more meaningful feature representation. LSTM has achieved great success in many NLP tasks but there have been some limitations. As these models are trained sequentially, it becomes really difficult to scale these models. Secondly, when we work with very long sequences, LSTMs suffer from context loss and thus they are not ideal for understanding the context of longer sequences. Due to some limitations, including the ones discussed here, new ways of learning context from sequential inputs were invented.

The advent of transformer-based models was groundbreaking for the field of NLP, as well as Vision AI. Transformer-based models heavily rely on attention mechanisms to capture the context and inter-sequence patterns, and they are also able to handle very long sequences of inputs. **Bidirectional Encoder Representations from Transformers (BERT)** is a family of NLP models that are based on some parts of the transformer architecture. BERT-based models have achieved great success in tons of NLP tasks, some of which seemed close to impossible in past decades.

Another advantage of working with deep learning models is that we don't have to train them from scratch every time. We always utilize pre-trained models and then fine-tune them on our domain-specific data to get great results faster and without requiring a lot of domain-specific training data. This approach is termed **transfer learning** and is where large deep learning models are pre-trained with huge amounts of data, after which they can be utilized for many downstream domain-specific tasks as they can be fine-tuned with a small amount of domain-specific training data.

## BERT for fake news classification

In this experiment, we will utilize a pre-trained BERT model and fine-tune it slightly on our news article training dataset. Let's get started.

## Importing useful libraries

In this experiment, we will utilize PyTorch as a framework for fine-tuning the BERT model. We also utilize the `transformers` library from Hugging Face to load the pre-trained weights of the BERT-based model with some other tooling that is useful for setting up fine-tuning:

```
import torch
from transformers import BertTokenizer
from transformers import BertForSequenceClassification
from transformers import AdamW
from transformers import get_linear_schedule_with_warmup
from torch.utils.data import TensorDataset
from torch.utils.data import random_split, DataLoader
```

Now, let's start preparing the dataset.

## The dataset

We will work with the same dataset that we used in the first experiment. So, we will follow the same steps here as well – we will load the data, treat NULL values, and create a content column with all the necessary text:

```
news_df = pd.read_csv("WELFake_Dataset.csv")
news_df.fillna('', inplace=True)
news_df['content'] = [x + ' ' + y for x,y in zip(news_df.title, news_
df.text)]
```

Next, we will convert the text in the content column to lowercase:

```
news_df['content'] = news_df['content'].apply(lambda text: text.
lower())
```

Now, we will separate text from labels and store them as lists:

```
texts = news_df.content.values
labels = news_df.label.values
print(len(texts), len(labels))
```

Here's the output:

72134 72134

Now, let's prepare our dataset as per the requirements of BERT model inputs.

## Data preparation

As we are working with the BERT model now, we don't need to perform lots of data cleanups, such as removing numbers, removing stopwords, stemming, and so on. Each BERT model has a tokenizer that is utilized to convert textual data into numeric IDs. So, we will need to find the appropriate BERT tokenizer (which can be loaded by the `transformers` library from Hugging Face), do tokenization, and also create attention masks for training purposes.

Let's create the tokenizer object:

```
tokenizer = BertTokenizer.from_pretrained('bert-base-uncased')
```

Here, we're defining a function that will create tokenized text and attention masks for training:

```
def prepare_tokenized_data(texts, labs='None'):
    global labels
    input_id_list = []
    attention_masks = []
```

Next, we must generate encodings and attention masks for each input text:

```
for text in tqdm_notebook(texts):
    encoded_dict = tokenizer.encode_plus(
        text,
        add_special_tokens = True,
        truncation = 'longest_first',
        max_length = 100,
        pad_to_max_length = True,
        return_attention_mask = True,
        return_tensors = 'pt'
    )
    input_id_list.append(encoded_dict['input_ids'])
    attention_masks.append(encoded_dict['attention_mask'])
```

Now, let's convert the lists into PyTorch tensors:

```
input_id_list = torch.cat(input_id_list, dim=0)
attention_masks = torch.cat(attention_masks, dim=0)

if labs != 'None':
    labels = torch.tensor(labels)
    return input_id_list, attention_masks, labels
else:
    return input_id_list, attention_masks
```

Here, we're calling the necessary method, to prepare our data:

```
input_id_list, attention_masks, labels = (
    prepare_tokenized_data(texts, labels)
)
```

Now, let's split our data into training and test partitions.

## Splitting the data

Here, we will create a tensor dataset out of our input IDs, attention masks, and labels and split it into training and test sets. Similar to the first experiment, we will be using about 80% of the data for training (or fine-tuning) purposes and the remaining 20% to test the results and metrics. The following snippet shows how to create and split the tensor dataset:

```
tensor_dataset = TensorDataset(input_id_list, attention_masks, labels)

# lets keep 80% articles for training and 20% for test
train_size = int(0.8 * len(tensor_dataset))
```

```
test_size = len(tensor_dataset) - train_size

train_data, test_data = random_split(tensor_dataset, [train_size,
test_size])
print(len(train_data.indices), len(test_data.indices))
```

Here's the output:

```
(57707, 14427)
```

Now, let's define data loader objects with the required batch size.

## Creating data loader objects for batching

The next step is to create data loader objects for both the training and test partitions. We will have a batch size of 32 for the training data and a batch size of 1 for the test data:

```
batch_size = 32
num_workers = 4

train_data_loader = DataLoader(
    dataset=train_data,
    batch_size=batch_size,
    shuffle=True,
    num_workers=num_workers,
)
# test data loader with batch size of 1
test_data_loader = DataLoader(
    dataset=test_data,
    batch_size=1,
    shuffle=False,
)
```

Our data is now ready for the model. This means we can load the model and start training.

## Loading the pre-trained BERT model

Here, we will load the pre-trained weights of the BERT-based model so that we can fine-tune it further on our custom dataset. The pre-trained weights of many BERT variants are available on Hugging Face and can be loaded through the transformers library, as shown in the following snippet:

```
device = 'cpu'
bert_model = BertForSequenceClassification.from_pretrained(
    'bert-base-uncased',
    num_labels=2,
```

```
        output_attentions=False,
        output_hidden_states=False,
    )
bert_model.to(device)
```

Using the `device` variable, we can choose to load our model on an accelerator, such as a GPU. This snippet downloads the pre-trained weights of the `bert-base-uncased` model with a classification layer of two labels. Executing this snippet also prints the BERT architecture summary, which will look something similar to the following:

```
BertForSequenceClassification(
    (bert): BertModel(
        (embeddings): BertEmbeddings(
            (word_embeddings): Embedding(30522, 768, padding_idx=0)
            (position_embeddings): Embedding(512, 768)
            (token_type_embeddings): Embedding(2, 768)
            (LayerNorm): LayerNorm((768,), eps=1e-12, elementwise_
affine=True)
            (dropout): Dropout(p=0.1, inplace=False)
        )
        . . . . . . . . . . .
        . . . . . . . . . . .
```

Now that our model has been loaded, let's define the optimization settings.

Optimizer

Here, we're defining the `AdamW` optimizer and setting a custom learning rate:

```
optimizer = torch.optim.AdamW(
    bert_model.parameters(),
    lr=6e-6,
    eps=1e-8,
)
```

Let's also define a scheduler for our model training.

## Scheduler

Here, we're setting up training steps and a training scheduler. We're planning to fine-tune our model for just 3 epochs on our training partition, after which we will check the results on our test set:

```
num_epochs = 3
steps_per_epoch = len(train_data_loader)
total_steps = steps_per_epoch * num_epochs
```

```
scheduler = get_linear_schedule_with_warmup(
    optimizer,
    num_warmup_steps = 0,
    num_training_steps = total_steps,
)
```

Now, we are all set to start training the model.

## Training BERT

Here, we will fine-tune the BERT model on our training data for 3 epochs, as defined in the previous sub-section:

```
bert_model.train()
for epoch in range(num_epochs):
    total_loss = 0
    for i, (ids, masks, labels) in enumerate(train_data_loader):
        ids = ids.to(device)
        masks = masks.to(device)
        labels = labels.to(device)
        loss = bert_model(ids, token_type_ids=None, attention_
mask=masks, labels=labels)[0]
        optimizer.zero_grad()
        loss.backward()
        optimizer.step()
        scheduler.step()
        total_loss += loss.item()
    print('Epoch: {}, Loss: {:.4f}'.format(epoch+1, total_loss /
steps_per_epoch))
```

This training snippet prints the model loss after each epoch of training is completed. The loss output from our experiment is as follows:

```
Epoch: 1, Loss: 0.0803
Epoch: 2, Loss: 0.0229
Epoch: 3, Loss: 0.0112
```

Now that our model training (or fine-tuning) is complete, we can save the model weights:

```
# save trained model locally
torch.save(bert_model.state_dict(), 'BERT.ckpt')
```

With that, we have successfully trained and saved the model. Now, let's move on to model evaluation.

## Loading model weights for evaluation

Here, we're loading our trained model weights for evaluation purposes:

```
bert_model.eval()
bert_model.load_state_dict(
    torch.load('BERT.ckpt'),
)
Output:
<All keys matched successfully>
```

Let's check the accuracy of the trained model on the test dataset.

## Calculating the accuracy of the test dataset

Now that our model has been trained and loaded for evaluation, we can make predictions on the test dataset and check its accuracy.

We will store the predictions in lists and also count the number of correct predictions in the following variables:

```
correct_predictions = 0
predictions = []
reals = []
```

Here, we're running model predictions on the test data:

```
for i, (ids, masks, labels) in enumerate(test_data_loader):
    ids = ids.to(device)
    masks = masks.to(device)
    labels = labels.to(device)
    bert_out = bert_model(ids, token_type_ids=None, attention_
mask=masks, labels=labels)[1]
    prediction = torch.max(bert_out, 1)[1][0].item()
    true_label = labels[0].item()
    correct_predictions += int(prediction == true_label)
    predictions.append(prediction)
    reals.append(true_label)
```

We can calculate the accuracy by dividing the correct predictions by the total predictions:

```
avg_correct_predictions = correct_predictions / len(test_data)
print('Accuracy: {:.4f}\n'.format(avg_correct_predictions))
```

The output of this cell shows that our model has about 99% accuracy on the test dataset. This is a huge improvement over the classic random forest model:

```
Accuracy: 0.9902
```

Finally, let's print the confusion matrix:

```
print(confusion_matrix(reals, predictions,))
```

Here is the output:

```
[[7025   53]
 [  88 7261]]
```

Now, we can generate a classification report for our model.

## Classification report

Finally, we will print the classification report of our experiment to understand the precision, recall, and F1 score of each class:

```
print(
    classification_report(
        reals,
        predictions,
        target_names=['Real', 'Fake'],
    )
)
```

Here's the output:

```
              precision    recall  f1-score   support

        Real       0.99      0.99      0.99      7078
        Fake       0.99      0.99      0.99      7349

    accuracy                           0.99     14427
   macro avg       0.99      0.99      0.99     14427
weighted avg       0.99      0.99      0.99     14427
```

The preceding output shows that our BERT-based classification model is extremely accurate with an accuracy of about 99%. Similarly, for each class, we have precision and recall scores of about 99%. This experiment showed that using a pre-trained deep learning model can enhance the accuracy of classification by a great margin.

# Summary

This chapter was about a real-world NLP use case for detecting fake news. In the current era of the internet, spreading fake news has become quite easy and it can be dangerous for the reputation of a person, society, organization, or political party. As we have seen in our experiments, ML classification can be used as a powerful tool for detecting fake news articles. Deep learning-based approaches can further improve the results of text classification use cases without requiring much fine-tuning data.

After reading this chapter, you should be confident about training and applying classification models on text classification use cases, similar to fake news detection. You should also have a good understanding of the cleaning and pre-processing steps that are needed to apply classical models, such as random forest, on text data. At this point, you should be able to launch large-scale ML experiments as Vertex AI training jobs. Finally, you should have a good understanding of how deep learning-based BERT models can be applied and fine-tuned for text classification use cases.

# Index

## A

**ad hoc transformation**
categorical data, handling  45
numeric data, handling  44
within Jupyter Notebooks  43
**AI techniques  165**
global explainability, versus
local explainability  165
tabular data techniques  168
text data techniques  170
**area under the ROC curve (ROC-AUC)  371**
**artificial intelligence**
(AI)  3, 49, 133, 163, 241, 304
**Artificial Neural Networks (ANNs)  207**
**Atomicity, Consistency, Isolation**
and Durability (ACID)  42
**attention mechanisms  171**
**AUC PR  85**
**AUC ROC  85**
**AutoML  68, 306**
**AutoML for tabular data**
SHAP-based explanation, using  173
**AutoML for Text Analysis  313**
classification  314
entity extraction  314
sentiment analysis  314

**AutoML Translation  309-312**
**AutoML Video Intelligence  308**
use cases  308

## B

**batch normalization  140**
**Batch Normalization (BN)  348**
**batch predictions  88**
obtaining, on Vertex AI  237-239
**Bayesian optimization  191**
**BERT-based fake news**
classification  374, 375
classification report  382
data loader objects,
creating for batching  378
data preparation  376, 377
dataset  376
data, splitting  377
model weights for evaluation, loading  381
pre-trained BERT model, loading  378, 379
scheduler  379, 380
test dataset accuracy, calculating  381
useful libraries, importing  375

**Bidirectional Encoder Representations from Transformers (BERT)  375**
training  380
using, for fake news classification  375
**BigQuery (BQ)  40, 42, 101**
benefits  42, 43
**BigQuery Data Transfer Service  40**
**BigQuery Machine Learning (BQML)  42, 66, 99**
advantages  100
automatic preprocessing  102
limitations  100, 102
manual preprocessing  102-112
used, for building machine learning models  113, 114
using, for feature transformation  102
using, in hyperparameter tuning  122-126
**black-and-white images**
converting, into color images  134-146
**boosted tree model**
creating  118-120
**BQML models**
boosted tree model, creating  118-120
creating  114
deep neural network model, creating  116-118
for inference  129, 130
hands-on exercise  131
importing  120
K-means model  121
linear regression models  114
logistic regression models  114
random forest model, creating  118-120
wide-and-deep model, creating  116-118
**BQML trained model**
evaluating  126-129
options  127
**business impact metrics  326**

# C

**categorical data**
challenges  45
handling  45
nominal categorical data  45
ordered categorical data  45
**categorical data, conversion methods**
embeddings  46
label encoding  45
one-hot encoding  45
**click-through rate (CTR)  325**
**Cloud Composer  213, 227**
used, for orchestrating ML workflows  227
versus Vertex AI Pipelines  232, 233
**Cloud Composer environment**
creating  227-232
**Cloud Data Fusion  46**
features  46
**Cloud Scheduler  213**
**Cloud Translation API  309**
**coefficients in linear models  170**
**collaborative filtering  324**
**color images**
black-and-white images, converting into  134-146
**command line**
using  39
**content-based filtering  324**
advantages  324
challenges  324
**continuous data  44**
**continuous improvement  312**
**Contract Parser  296**
**convolutional neural networks (CNNs)  139, 167, 208, 347**

coverage metrics  327
CREATE MODEL statement
  key options  114-116
custom containers
  for Vertex AI Workbench  58
custom Document AI processors
  creating  300-302
customer lifetime value (CLV)  326
custom ML model  306
  AutoML  306
  custom training  307
custom training  307

## D

DART booster  119
data
  moving, into Google Cloud  38
data drift  244
Dataflow Flexible Resource
    Scheduling (FlexRS)  47
Dataflow pipelines
  features  47
  for scalable data transformation  47
data governance  242
data storage  41
  BigQuery (BQ)  42, 43
  Google Cloud Storage (GCS)  41
data transformation  43
  ad hoc transformation,
    within Jupyter Notebooks  43
  Cloud Data Fusion  46
decoder  140
deep learning (DL)  49
  building  330, 331
  building, with TensorFlow  134
deep neural network (DNN)  187
  creating  116-118

demographic filtering  325
  advantages  325
  challenges  325
deployed models
  managing, on Vertex AI  239
directed acyclic graph (DAG)  214
Document AI  293, 294
  key features  294
Document AI processors  294
  Contract Parser  296
  custom  295
  Document OCR  295
  Document Splitter  296
  Form Parser  295
  France Driver License Parser  296
  general  294
  Intelligent Document Quality Processor  296
  Invoice Parser  296
  Pay Slip Parser  296
  specialized  295
  using  296-300
  US Passport Parser  296
Document OCR  295
Document Splitter  296
document translation  312
drift detection  255
dropout layer  140

## E

Electronic Program Guides (EPGs)  319
embeddings  46
encoder  140
engagement metrics  326
event  250
execution  250
Explainable AI (XAI)  163, 164

Explainable AI (XAI) for MLOps
      practitioners  164
  ethical considerations  165
  model debugging and improvement  165
  regulatory compliance  164
  trust and confidence, building  164
exploratory data analysis (EDA)  50
eXtended Relevance-weighted Attribution
      of Importance (XRAI)  166
extractive question answering (EQA)  267

F

fake news, detecting with NLP
  classification, with random forest  364
  confusion matrix  371, 372
  data, cleaning  366-368
  data, pre-processing  366-368
  data, reading  365
  data, separating from label  368
  data, verifying  365
  dataset  364
  data, splitting  369
  model, training  369
  NULL value check  366
  random forest classifier, defining  369
  results/metrics, checking
      on test dataset  370, 371
  test data, predicting  369
  text, converting into numeric data  368
  title and text, combining
      into single column  366
  useful libraries, importing  365
false positive rate (FPR)  370
Fast Fourier Transformation (FFT)  43
feature-based explanation  172
feature importance  170
  GINI importance  170
  linear models coefficients  170

Form Parser  295
France Driver License Parser  296

G

GCP solutions, for model creation
  Big Query ML (BQML)  66
  Vertex AI AutoML  66
  Vertex AI custom training  66
GenAI applications
  building, and deploying with Vertex AI  282
GenAI models
  applications  261
  limitations, of standard entity
      extraction approach  283
  QA solution based on RAG
      methodology, implementing  283
  using, to extract key entities
      from scanned documents  282, 283
GenAI models, types
  autoregressive models  262
  generative adversarial networks (GANs)  262
  variational autoencoders (VAEs)  262
GenAI, with Vertex AI  265
  foundation models  265
  prompt  271
Generative Adversarial Networks
      (GANs)  146
Generative artificial intelligence
      (GenAI)  261
  challenges  262, 263
  fundamentals  262
  LLM evaluation  264
  model, types  262
  performance, enhancing
      with model tuning  286, 287
  versus traditional AI  262
  with Vertex AI  265

GINI index  170
global explainability  165
  versus local explainability  165
Google Cloud
  data, moving into  38
  ML modeling options  66, 68
  Natural Language AI  313
  Speech AI  317
  Translation AI  308
  Vision AI  304
Google Cloud Console UI
  using  39
Google Cloud Platform Billing
  reference link  257
Google Cloud Platform
    (GCP)  22, 38, 66, 101, 257
Google Cloud Storage (GCS)  38, 41, 333
  benefits  41
  limitations  42
  vision model, saving  355
Google Cloud Storage Transfer tools  38
  command line, using  39
  Google Cloud Console UI, using  39
  REST API (JSON API)  40
Google Cloud Vertex AI  172
  aspects  24
  example-based explanation  172, 186, 187
  exercise  173-187
  feature-based explanation  172
  features  172
  SHAP-based explanation,
      with AutoML for tabular data  173
  used, for implementing MLOps  24-26
Google Compute Engine (GCE)  58
Google Container Registry (GCR)  204
Gradient Boosting Tree (GBTREE)  119
Gradient-weighted Class Activation
    Mapping (Grad-CAM)  167

graphical processing units (GPUs)  12, 355
grid search  191

H

Healthcare Natural Language API  315-317
  features  316
HPT jobs
  setting up, on Vertex AI  191-206
hyperparameters  190
hyperparameter tuning
  with BQML  122, 123, 126
Hyperparameter Tuning (HPT)  190
  advantages  190
  search algorithms, running  190

I

Identity and Access Management
    (IAM)  18, 40, 251
image data techniques  166
  eXtended Relevance-weighted Attribution
    of Importance (XRAI)  166
  Gradient-weighted Class Activation
    Mapping (Grad-CAM)  167
  Integrated Gradients  166
  Local Interpretable Model-agnostic
    Explanations (LIME)  167
Integrated Gradients  166
Intelligent Document Quality Processor  296
Interactive Voice Response (IVR)  317
Internet of Things (IoT)  317
Invoice Parser  296

J

Jupyter Notebooks  50-52

## K

key performance indicators (KPIs)  325
K-means model  121
Kubeflow  214

## L

label encoding  45
large language models (LLMs)  12, 262
  approaches, for customizing  287
  tunning  286
LeakyReLU activation  140
linear regression models  114
LLM evaluation methods
  benchmarking  264
  domain-specific benchmarks  264
  generative metrics  264
  human evaluation  264
local explainability  165
  versus global explainability  165
Local Interpretable Model-agnostic
    Explanations (LIME)  167, 168
logistic regression models  114
Long-Short-Term-Memory (LSTM)  375

## M

machine learning (ML)  3, 49, 133, 163, 241
  limitations  12
  building, with BQML  113, 114
machine learning operations
    (MLOps)  17, 241, 242
  bias, limiting in AI solutions  243
  costs, monitoring  244
  data governance  242
  enterprise scenarios  243
  implementing, with Google
      Cloud Vertex AI  24-26

  model governance  242, 243
  need for  18
  shifts, monitoring in feature
      distribution  243, 244
  training data, monitoring  244
Machine Learning Project/Development
    Lifecycle(MDLC)  4
manual preprocessing  103-112
Matching Engine  286
Mean Squared Error (MSE)  143
Medical Transcription Dataset
  running  288
MetadataSchema  250
Metadata Store  247-249
  collaboration and communication  248
  compliance and regulatory adherence  248
  consistency and standardization  248
  key concepts and terminology  250
  model experimentation and comparison  248
  reproducibility  248
  traceability  248
ML limitations  13
  cost concerns  14
  customization concerns  14
  data-related concerns  13
  deterministic nature of problems  13
  ethical concerns and bias  15
  lack of interpretability  14
  lack of reproducibility  14
MLOps maturity level 0  19, 20
MLOps maturity level 1  20, 21
MLOps maturity level 2  22, 23
MLOps maturity levels
  implementing  19
  MLOps maturity level 0  19, 20
  MLOps maturity level 1  20, 21
  MLOps maturity level 2  22, 23

**ML pipeline**
  components 21, 22
**ML project life cycle 3, 4**
  consensus, on results 7, 8
  data analysis 5
  feature engineering 6
  ML use case definition 4
  model design iterative process 6
  model operationalization 8
  model type selection 5, 6
  monitoring 9
  retraining 9
**ML workflows**
  orchestrating, with Cloud Composer 227
  orchestrating, with Vertex AI Pipelines 214
**model governance 242**
  model auditing 242
  model deployment management 242
  model monitoring 243
**Model Registry 245**
  centralized repository for models 246
  model metadata management 246
  model validation and testing 246
  version control and model lineage 246
  Vertex AI services, integrating 247
**model training, launching on Vertex AI**
  configurations, setting 373
  job, running 374
  training job, defining 373
  Vertex AI SDK, initializing 373
**model training progress**
  monitoring 156-160
**model tuning**
  GenAI performance, enhancing 286, 287
  safety filters, for generated content 289-291
  Vertex AI supervised tuning, using 287-289

**MovieLens dataset 328**
**movie recommender system, on Vertex AI**
  data preparation 328, 329
  deploying 327, 328
  local model testing 331, 332
  model building 330, 331
  model, deploying on Google Cloud 333, 334
  model for inference, using 334-338
**Multiple Additive Regression Trees 119**

# N

**Natural Language AI**
  AutoML for Text Analysis 313
  Healthcare Natural Language API 315-317
  Natural Language API 314, 315
  on Google Cloud 313
**Natural Language API 314, 315**
  use cases 314
**natural language inference (NLI) 261**
**natural language processing (NLP) 293**
  fake news, detecting with 364
**net promoter score (NPS) 326**
**Neural Architecture Search (NAS) 208**
  best practices 210
  components 208
  evaluation method 208
  launching, with Vertex AI 209, 210
  optimization method 208
  search space 208
**neural machine translation (NMT) 309**
**numeric data**
  bucketing 44
  handling 44
  normalizing 44

# O

**objectives, for classification problems**
AUC PR  81
AUC ROC  81
log loss  81
precision at recall  81
recall at precision  81
**objectives, for regression problems**
MAE  82
RMSE  82
RMSLE  82
**one-hot encoding  45**
**online predictions  88**
obtaining, from vision model  359-361
obtaining, on Vertex AI  234-237
**optical character recognition (OCR)  282, 293**
**orchestration  213**
**Oxford-IIIT Pet dataset  135**

# P

**page-based pricing  313**
**parameter-efficient fine-tuning (PEFT)  287**
**partial dependence plots (PDP)  170**
**Pay Slip Parser  296**
**permutation feature importance  170**
**Personal Identifying Information (PII)  10**
**pgvector extension  286**
**precision-recall (PR)  85**
**predictions**
generating  92, 93
generating, programmatically  93
generating, with batch predictions  88
obtaining, on Vertex AI  233
requests, submitting with REST API  93-97

**prompt  271, 287**
content  272-275
**prompt design**
aspects  272
**prompt engineering  271**
**Python**
used, for developing Vertex
AI Pipeline  214, 216

# Q

**queries per second (QPS)  240**

# R

**random forest model**
creating  118-120
**random search  191**
**real-world ML project, challenges  9, 10**
data collection  10
infrastructure requirements  12
non-representative training data  10
overfitting training dataset  11, 12
poor quality data  11
underfitting training dataset  11
**Recall-Oriented Understudy for Gisting Evaluation (ROUGE)  264**
**receiver operating characteristic (ROC)  85**
**recommender systems**
metrics, types  326, 327
real-world evaluation  325-327
types  324, 325
**Redshift  40**
**reinforcement learning (RL)  208, 287**
**ReLU  348**
**REST API (JSON API)  40**

**Retrieval Augmented Generation (RAG)  283**
  advantages  284, 285
  working  283-285
**revenue per user (RPU)  327**

# S

**scalable data transformation**
  using, in Dataflow pipelines  47
**scikit-learn  102**
**search algorithms**
  Bayesian optimization  191
  grid search  191
  random search  191
**self-serve translation  312**
**sentiment analysis (SA)  287**
**SHAP-based explanation**
  with AutoML for tabular data  173
**SHapley Additive exPlanations (SHAP)  169**
**simplified administration  312**
**skew detection  255**
**Software-as-a-Service (SaaS)  40**
**software development kit (SDK)  26**
**Software Development Life Cycle (SDLC)  4**
**Speech AI**
  on Google Cloud  317
  Speech-to-Text  317
  Text-to-Speech  318, 319
**Speech-to-Text  317**
  features  318
  use cases  317
**Storage Transfer Service  40**
**supervised learning (SL)  262**
**supervised tuning job**
  running  287

# T

**tabular data**
  used, for creating Vertex AI AutoML  70
**tabular data techniques  168**
**TensorFlow  102, 134**
  used, for building deep learning model  134
**TensorFlow Data Validation (TFDV)  255**
**TensorFlow Extended (TFX)  28**
**TensorFlow Extended (TFX)-
  based ML pipelines  214**
**TensorFlow (TF)-based deep
  learning model  343**
**Tensor Processing Units (TPUs)  12, 355**
**Teradata  40**
**text data techniques  170**
**text-specific LIME  170, 171**
**text-specific SHAP  171**
**Text-to-Speech  318, 319**
  features  319
  use cases  319
**TF model**
  uploading, to Vertex Model
    Registry  356, 357
**TF model architecture  347**
  convolutional block  348
  definition  348, 349
**Total Cost of Ownership (TCO)  257**
**trained models**
  evaluating  160, 161
**Transfer Appliance  40**
**transfer learning  375**
**Translation AI**
  AutoML Translation  309-312
  Cloud Translation API  309
  on Google Cloud  308
  Translation Hub  312

Translation Hub 312
  continuous improvement 312
  document translation 312
  page-based pricing 313
  self-serve translation 312
  simplified administration 312
true positive rate (TPR) 370

# U

unsupervised learning (UL) 262
user satisfaction metrics 326
US Passport Parser 296

# V

Vertex AI
  batch predictions, obtaining on 237-239
  deployed models, managing on 239
  HPT jobs, setting up 191-206
  model, deploying 88, 90, 92
  model package, submitting as
    training job 147-156
  model training, launching on 372, 373
  multiple models, deploying
    to single endpoint 239
  notebook executions, configuring 59-63
  notebooks, scheduling 59
  online predictions, obtaining on 234-237
  predictions, obtaining on 233
  resources, computing 240
  scaling 240
  single model, deploying
    to multiple endpoints 239
  used, for launching NAS 209

Vertex AI AutoML 66, 69
  creating, with tabular data 70
  data, importing to use with 70-75
  trained model, evaluating 83-87
  training, for tabular/structured data 76-82
Vertex AI Batch Predictions 32
Vertex AI Data Management 26
Vertex AI endpoint
  creating 357, 358
  vision model, deploying to 355-359
Vertex AI Endpoints 32
  features 32
Vertex AI Experiments 31
Vertex AI Feature Store 27, 28
  features 27
  using, to catalog and monitor
    features 253-255
Vertex AI foundation models
  chat mode 270
  prompt mode 270
Vertex AI GenAI models
  using, through GenAI Studio 275
Vertex AI GenAI models, use cases
  code samples, generating 281
  images, generating with GenAI
    Studio (Vision) 280, 281
  text, generating with free-form input 276
  text, generating with structured
    input 277-279
Vertex AI GenAI Studio 270
  Language section 270
Vertex AI GenAI toolset
  Vertex AI GenAI Studio 270
  Vertex AI Model Garden 265, 266
Vertex AI Metadata Store
  using, to track ML model 251, 252

**Vertex AI Model Garden  265, 266**
  foundation GenAI models  267-270
**Vertex AI Model Registry  30**
  features  30
**Vertex AI Monitoring  33**
  best practices  256
  features  33
  using, to track deployed models performance
    in production environments  256
**Vertex AI Pipeline  28**
  components, defining  216-227
  data loading component  217
  developing, with Python  214, 216
  features  28
  model deploying component  217
  model evaluation component  217
  model training component  217
  used, for orchestrating ML workflows  214
  versus Cloud Composer  232, 233
**Vertex AI SDK  214**
  installing  373
**Vertex AI supervised tuning**
  using  287-289
**Vertex AI tools  245**
  billing monitoring  257
  Metadata Store  247-250
  Model Registry  245
**Vertex AI Training  31**
  key capabilities  31
**Vertex AI UI  214**
**Vertex AI Vision  304-306**
**Vertex AI Workbench  26, 52**
  custom containers  58
  implementing  53-57
  key features  52
**Vertex ML Metadata Store  29**
  key parameters  29

**Vertex Model Registry**
  TF model, uploading  356, 357
**Video AI  307**
  AutoML Video Intelligence  308
  use cases  307
**virtual machine (VM)  26**
**Vision AI  304**
  custom ML model  306
  on Google Cloud  304
  Vertex AI Vision  304-306
  Vision API  307
**Vision API  307**
**vision-based defect detection  340**
  classification report  354, 355
  data, loading and verifying  341
  data preparation  343-345
  dataset  340
  data, preparation  346, 347
  data, splitting into train and test  345, 346
  data, testing  346, 347
  data, training  346, 347
  model, compiling  350
  model, training  350
  results  352
  results, checking on few random
    test images  353, 354
  samples, checking  341-343
  TF model architecture  347
  training progress, plotting  351, 352
  useful libraries, importing  341
**vision model**
  deploying, to Vertex AI endpoint  355-359
  online predictions, obtaining  359-361
  saving, to Google Cloud Storage (GCS)  355

# W

**wide-and-deep model**
  creating  116-118
**Workflows  213**

# X

**XAI techniques, for tabular data**
  feature importance  170
  Local Interpretable Model-agnostic
    Explanations (LIME)  168
  partial dependence plots (PDP)  170
  permutation feature importance  170
  SHapley Additive exPlanations (SHAP)  169
**XAI techniques, for text data**
  attention mechanisms  171
  text-specific LIME  170, 171
  text-specific SHAP  171

Packtpub.com

Subscribe to our online digital library for full access to over 7,000 books and videos, as well as industry leading tools to help you plan your personal development and advance your career. For more information, please visit our website.

## Why subscribe?

- Spend less time learning and more time coding with practical eBooks and Videos from over 4,000 industry professionals

- Improve your learning with Skill Plans built especially for you

- Get a free eBook or video every month

- Fully searchable for easy access to vital information

- Copy and paste, print, and bookmark content

Did you know that Packt offers eBook versions of every book published, with PDF and ePub files available? You can upgrade to the eBook version at packtpub.com and as a print book customer, you are entitled to a discount on the eBook copy. Get in touch with us at customercare@packtpub.com for more details.

At www.packtpub.com, you can also read a collection of free technical articles, sign up for a range of free newsletters, and receive exclusive discounts and offers on Packt books and eBooks.

# Other Books You May Enjoy

If you enjoyed this book, you may be interested in these other books by Packt:

**Journey to Become a Google Cloud Machine Learning Engineer**

Dr. Logan Song

ISBN: 978-1-80323-372-7

- Provision Google Cloud services related to data science and machine learning
- Program with the Python programming language and data science libraries
- Understand machine learning concepts and model development processes
- Explore deep learning concepts and neural networks
- Build, train, and deploy ML models with Google BigQuery ML, Keras, and Google Cloud Vertex AI
- Discover the Google Cloud ML Application Programming Interface (API)
- Prepare to achieve Google Cloud Professional Machine Learning Engineer certification

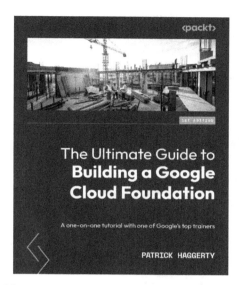

**The Ultimate Guide to Building a Google Cloud Foundation**

Patrick Haggerty

ISBN: 978-1-80324-085-5

- Create an organizational resource hierarchy in Google Cloud
- Configure user access, permissions, and key Google Cloud Platform (GCP) security groups
- Construct well thought out, scalable, and secure virtual networks
- Stay informed about the latest logging and monitoring best practices
- Leverage Terraform infrastructure as code automation to eliminate toil
- Limit access with IAM policy bindings and organizational policies
- Implement Google's secure foundation blueprint

## Packt is searching for authors like you

If you're interested in becoming an author for Packt, please visit authors.packtpub.com and apply today. We have worked with thousands of developers and tech professionals, just like you, to help them share their insight with the global tech community. You can make a general application, apply for a specific hot topic that we are recruiting an author for, or submit your own idea.

## Share Your Thoughts

Now you've finished *The Definitive Guide to Google Vertex AI*, we'd love to hear your thoughts! Scan the QR code below to go straight to the Amazon review page for this book and share your feedback or leave a review on the site that you purchased it from.

https://packt.link/r/1-801-81526-7

Your review is important to us and the tech community and will help us make sure we're delivering excellent quality content.

# Download a free PDF copy of this book

Thanks for purchasing this book!

Do you like to read on the go but are unable to carry your print books everywhere? Is your eBook purchase not compatible with the device of your choice?

Don't worry, now with every Packt book you get a DRM-free PDF version of that book at no cost.

Read anywhere, any place, on any device. Search, copy, and paste code from your favorite technical books directly into your application.

The perks don't stop there, you can get exclusive access to discounts, newsletters, and great free content in your inbox daily

Follow these simple steps to get the benefits:

1. Scan the QR code or visit the link below

https://packt.link/free-ebook/978-1-80181-526-0

2. Submit your proof of purchase
3. That's it! We'll send your free PDF and other benefits to your email directly